T0181893

Elementary Statistical Methods

Elementary Statistical Methods

SAHANA PRASAD

Sahana Prasad

Elementary Statistical Methods

Sahana Prasad
Christ University
Bangalore, India

ISBN 978-981-19-0598-8 ISBN 978-981-19-0596-4 (eBook)
https://doi.org/10.1007/978-981-19-0596-4

This Springer imprint is published by the registered company Springer Nature Singapore Pte Ltd.
The registered company address is: 152 Beach Road, #21-01/04 Gateway East, Singapore 189721, Singapore

MISS TICS

Hi, I am Miss TICS and I always emphasise learning with examples and encourage the application of these concepts in reality. Let's code, and decode the inner meaning of data together. Let's hunt for relevant programming tools to solve the mystery of numbers. With this, you can get hands-on experience in solving every problem with a statistical solution, meaningful interpretation and a reliable conclusion. Let's get Sherlocked with curious case studies!

Thoughtful

Interesting examples with

Case

Studies

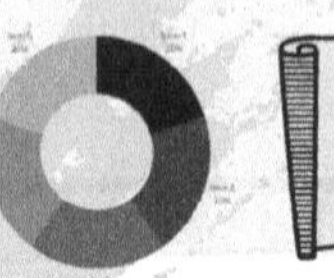

MR STAT

Hi, I am Mr STAT and I believe in lucid learning. In this book, I have simplified all the complex concepts into manageable chunks of theory in a narrative manner. I have also added plenty of interesting charts, meaningful graphs, thoughtful tables and also a few data comics!
All these will help you have a clear understanding of statistical concepts with an analytical bent of mind!

Simplistic

Theories with

Accurate

Trendy graphs

Let's learn statistics together !
Simple Theory
Meaningful graphs
Data comics
Thoughtful tables
Interesting charts
Real life examples
Practical Problems
Curious case studies
Code + Math
Mr. STAT
Miss TICS
Stat IS Tics
Always
With

Preface

Statistics is basically a mathematical science wherein data is first collected, then analysed by using suitable tools, explanation/interpretation given, and then finally converted into a valid conclusion. Statistical techniques are used right from data collection to final measurement. Different experimental designs and survey sampling techniques help in obtaining the data while data quality is improvised by statisticians by using data cleaning techniques. Various forecasting and prediction tools are available for providing inferences. Statistical knowledge is a relevant discipline in various academic organizations—in social and natural sciences and in business set-ups and governmental organizations.

The world is producing more statistical information than ever before, and many people feel overwhelmed by it. It is important to know what information we should pay attention to what to do with it and how to use them. Statistics are information but, as Albert Einstein put it, "information is not knowledge". Yet, it is the knowledge that leads to good decision-making and spurs progress. Statistics are raw material for the creation of knowledge, and obviously, the quality of statistics is critical for public policy. Flawed analysis affects conclusions and can lead to poor policy decisions, thus negating the purpose of data analysis. As Benjamin Franklin said, "an investment in knowledge pays the best interest".

A well-planned program is essential before collecting the data. Knowledge about alternative statistical methods, their assumptions, limitations and their usability is very important to the researcher. Many times, a researcher may not be able to determine the appropriate tool to use for different kinds of data. He/she may apply a rigid theoretical approach which may not be correct. When standard questions are used, it may lead to structural bias and false representation. They may also collect a narrow and superficial data set, and the results can get distorted because of false assumptions and inaccurate use of techniques.

Since political, economic and other social conditions are rapidly changing, one has to be equipped to deal with them and use the appropriate tools with caution. This requires that the researcher has enough exposure and familiarity with different tools and techniques as the onus of selecting and implementing the right one rests with the researcher. In order to fulfil this responsibility, the researcher has to pay attention to every small details of the study, right from data collection, measurement of data, by using the right tool for different types of data and making the correct inferences from the data by using the correct and valid techniques of analysis. A very important aspect of this is that the researcher should become conversant with the available tools and the situations in which they can be used. Thus, he is like a skilled doctor who knows exactly what has to be prescribed and to whom.

Statistics aids the researcher in giving valuable and concrete support in terms of tools and theory. The most important point to be noted is that experiments/researches have to be designed to use statistics. Many researchers embark on a study without pausing to think of how they can use statistics right from the design stage. If an experiment is designed and executed properly, as we would expect of an eminent scientist, then the results often speak for themselves. A good experimental design involves having a clear idea about how we will analyse the results when we get them. That's why statisticians often tell us to "think about the statistical tests we will use before we start an experiment".

A researcher in any field will surely benefit by taking up a course in basic statistics to know the usage and appropriateness of various concepts. Depending on software for analysis and, thus on experts in that area may not lead to a good analysis, if the theory behind the problem has not been understood. It has been found that many research scholars use various measures without paying attention to data, its scale and its validity. This may produce some results, but it may turn out to be a skewed analysis. A wrong tool may lead to erroneous conclusions, which may affect the credibility of the researcher. It is a good principle, therefore, to remember that "a tool is good in the hands of a skilled and knowledgeable researcher but may turn out to be a dangerous weapon when used by an amateur".

There are a large number of user-friendly packages available to aid research. SPSS (renamed PASW) STATA, SAS, EXCEL, R, etc., are some other software which are used to analyse data. However, usage of these requires caution as they operate like other software packages on the principle of "garbage-in, garbage-out". Thus, it is essential to use the right statistical tools and to have at least a basic knowledge of statistics before using software.

This book deals with the above issues with reference to statistical tools. It assumes that the reader has very elementary or no knowledge about statistical tools and tries to explain all concepts by using case studies. There is just enough theory to introduce the concept and then take the reader through a variety of case studies. These answer four main questions, namely *what*, *why*, *how* and *when*. All concepts are explained keeping these questions, which are likely to arise while understanding statistical concepts.

A detailed list of figures as well as tables is appended in the book. The links given in the book are available in https://github.com/ElementaryStatisticalMethods/Book1.git.

Bangalore, India Sahana Prasad

Acknowledgements

I would like to thank the Almighty who has been instrumental in guiding me in envisaging and creating this book. His unseen hand has led me up the path of writing a useful and simple book.

I thank Springer Nature and the Executive Editor Shamim Ahmed in particular, who encouraged and supported me in this arduous journey. I am grateful to them as this book could not see the light of the day, otherwise.

My heartfelt thanks to my husband Mr. L. V. Prasad for his support and encouragement. My daughter Sharanya, son Amit Vikram, my son-in-law Vineeth Patil and most importantly, my dear grandson Rihaan who have stood by me and lent their suggestions and time in shaping this book. I am also thankful to my father B. N. Vishwanath who has always encouraged me in all my endeavours.

Ms. Srishma Sunku, my collaborator, has helped me immensely by proofreading, creating infographics, adding new content and editing the book thoroughly. I am indebted to her as the book has been improved widely by her efforts. She is the editor who has verified the contents as well as added the infographics, wherever needed.

My students, colleagues, fellow researchers and other coworkers whose suggestions and requirements have dictated the content of this book need special mention. I thank each and every one of them profusely.

Sahana Prasad

Contents

About the Author

SAHANA PRASAD is former Associate Professor at the Department of Statistics, Christ University, Bangalore, India. She holds a Ph.D. degree in applying operations from Christ University, Bangalore, India. She has taught various topics in statistics, operations research, quantitative techniques, and data science to students of courses like chartered accountancy, company secretaryship, commerce, economics, management, among others. She also has been a resource person for many training programs and given guest/keynote lectures at several conferences and seminars. She has conducted workshops in SPSS, R, and EXCEL and has a working knowledge of MATLAB, LINGO, TABLEAU, and PYTHON. An awardee of the Statistical Capacity Building award by the World Bank, Dr. Prasad was featured in Asia Research news for her work in statistics. She has around 20 research publications to her name and around 350 articles/short stories published in various newspapers and magazines.

List of Figures

List of Tables

Chapter 1
Basic Concepts in Research and Data Analysis

WHAT

Data and its types, scales of measurement, tools for data representation.

WHY

- It is very important to comprehend the characteristics of the data we are dealing with.
- To locate the influencers of the data and comprehend the magnitude of these influencers.
- To translate large data into simple visualisations that are impactful to the readers.

HOW

By exploring the data through descriptive statistical tools like tabulation and visualisations of data.

WHEN

- Curating data into useful information.
- The concepts defined in this chapter are used in the first step of every data analysis.

S. Prasad, *Elementary Statistical Methods*,
https://doi.org/10.1007/978-981-19-0596-4_1

Basic concepts in research and data analysis

1.1 Introduction: Understanding Data and a Few Terminologies

In real life, we come across a lot of information. When analysed thoughtfully help us make informed decisions. We can use fundamentals of math, logical tools and statistical methodologies to decode the nuances of the data.

For example,

A CEO of a company is keen on figuring out the happiness index of his employees. In such a case, employee happiness has to be defined precisely.

1. Are you referring to their mental state when they work in an office?
2. Their mental state when they get back home after work?
3. Or their general perception of the workplace?

Once the unknowns are clearly defined, we have resolved 50% of our problem. The next important step is to choose the right path towards solving this problem.

1. What, when, how and where do I find the information?
2. What are the tools/techniques used to solve this problem?
3. Am I logical in my thinking? (Fig. 1.1)

Fig. 1.1 The path towards solving a data problem

What Exactly Do We Mean by "Data"?

Merriam

Almost all research involves a process in which a researcher takes a general concept or idea, specifies the dimensions that he/she wants to study, and then creates measures to evaluate these dimensions. These processes are known as **conceptualization and operationalization.** Singleton and Straits refer to the process of establishing a conceptual definition in addressing the first of these questions while addressing the second is the creation of an operational definition.

Data (plural of datum) can be defined as the lowest unit of information in the form of text, symbols, diagrams, graphs, numbers and many more.

1. **Data**: Raw/unprocessed information. It is of two types:

   (a) **Qualitative Data**: A type of data that cannot be numerically measured, also called ***categorical*** as it reveals the characteristic features. Example: shapes of candies, colours of cars, your weekend plans, etc. If an employer observes an employee's behaviour, in terms of dedication, concentration, sincerity, etc., this constitutes qualitative data and the characteristics observed are termed as *attributes*. Using such data, we can find the number or frequency of a particular trait or characteristic. For example, a company in the hospitality industry will be interested in customer feedback on the issues they are facing, the satisfaction of service provided and so on. All emotions and perceptions fall under the category.

Qualitative data can be classified as follows:

- **Unifold classification**: When classification is done based on only one attribute. For example, a hospital management system categorizes patients in the hospital, based on their type of sickness.
- **Bifold classification**: When classification is done based on two attributes. For example, a hospital management system categorizes the patients in the hospital, based on their types of sickness and the type of treatment provided to them.
- **Manifold classification**: When classification is done based on three or more attributes. A hospital management system categorizes the patients in the hospital, based on their type of sickness, the type of treatment provided to them and their response to the treatments (Fig. 1.2).

Quantitative Data: A type of data that can be quantified/measured. Also called ***numerical data***. And the characteristics measured are called variables. Example: profit and loss of a financial institution, fuel prices in an economy, student's examination scores, weights of iron in furnaces and many more. There are two types of quantitative data:

- **Discrete Data**: Data that is in the form of finite countable units. They can take only certain values in a given range. For example, the number of pea pods in a pea plant, the number of errors made by a typist and many more (Fig. 1.3).

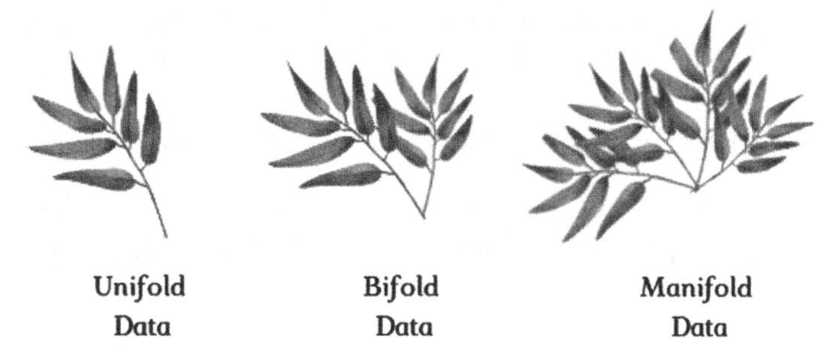

Fig. 1.2 Unifold, bifold and manifold data

Fig. 1.3 Types of data

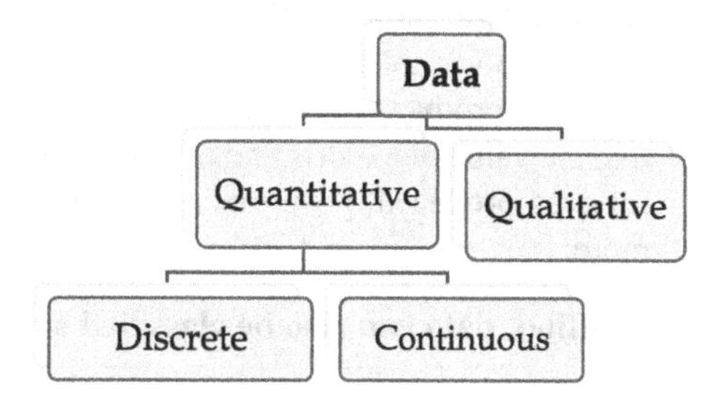

- **Continuous Data**: Data that is defined over a range and can take any value in that range. It is written in the form of class intervals. Examples: temperatures, wind speed, income, etc.

Quantitative data can also be classified as follows:

- **Univariate data**: "Uni" refers to one and variate refers to a variable. Therefore, as the name suggests, classification based on one variable is called the univariate classification of data. For example, favourite books of kids, sections of government policies, types of rice grains produced, etc. This type of classification is suitable for nominal and categorical data. At times when analysis gets complicated with too many factors to consider, the analyst will evaluate each factor independently using univariate analysis.
- **Bivariate data**: Bi refers to two. Therefore, classification based on two variables is called bivariate classification of data. For example, sales of red roses on the eve of every New Year, seasons and temperatures, demand and supply of goods, etc. Most of our descriptive data analysis will use bivariate data. All the dimensional graphs plotted on the x and y-axis are examples of bivariate data.

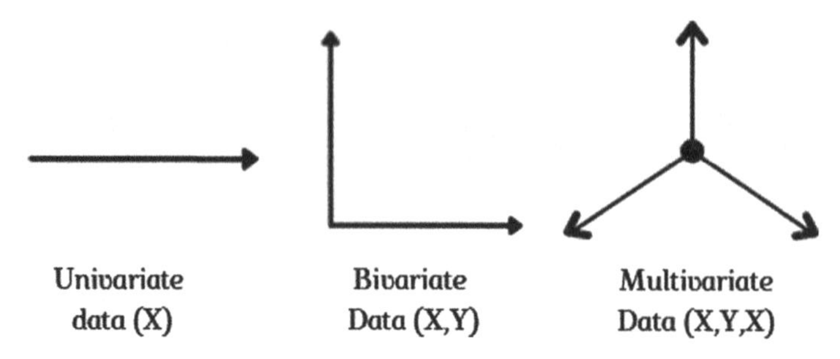

Fig. 1.4 Univariate, bivariate and multivariate data classification

- **Multivariate data**: Multi refers to many and thus classification based on more than three variables is called multivariate classification. For example, the gross domestic product (GDP) of the country has many variables under consideration, like the value of exports and imports, total income generated from the goods and services produced, government spending, investments, foreign loans and many more.

In addition, data can also be classified as grouped data and ungrouped or raw data.

- **Grouped data**: When we classify and categorize data into various groups or mention the various values of data in any form, ascending or descending, it is known as grouped data. These include both discrete and continuous data.
- **Ungrouped data**: Refers to individual raw data which is not grouped or classified in any manner. Collected data is raw data, till it is classified (Fig. 1.4).

Frequency tables and their components: When we represent the values of a variable along with the number of times it occurs, we get a frequency distribution of the variable. It has the following components:

(a) **Value of the variable/class intervals**: The various values or the intervals in which the variable can take/lie.
(b) **Tally Marks**: These are defined under the unary number system. A tally mark is a simple vertical line drawn. One line indicates one count. [|—one], [||—two], [|||—Three] and [||||—four] and the fifth line is struck across the four as shown in the icon. This way counting becomes simple. Tally marks are used when we need to score data quickly, these hash marks are much faster than writing numbers. Tally marks can be eliminated when we calculate the final count.
(c) **Frequency**: This refers to the number of times an event is repeated in a given period. Usually denoted by "**f**".
(d) **Class Intervals**: The range of values which a variable can take is given by class intervals. 400–500 kg, 3.9–6.5 centimetres, etc., is called a **class interval or a class of data.** The two endpoints of this interval are called the **upper and lower limit** respectively. The width of the class interval is simply the difference

in the upper and lower limits. There are three types of class intervals which are as follows:

1. **Inclusive**: When both the lower and the upper limit are included in the interval it is called an inclusive class interval. For example: 10–14, 15–19, 20–24. This means, data point 14 and 10 are both included in the interval mentioned.
2. **Exclusive**: When the lower limit is included and the upper limit is excluded from that class interval but included in the successive class interval, this is called exclusive type. Example: 10–15, 15–20, 20–25. This means if 15 has to be categorized, then it is excluded from the interval which is 10–15 but it is included in the next class interval which is 15–20.

   NOTE: Almost all statistical formulae require that the class intervals should be in exclusive form. Only mean and standard deviation will work for inclusive as well as exclusive class intervals. It is always a good practice to construct exclusive class intervals unless it is specifically required for research. Also, inclusive class intervals give no scope to enter certain values. Example: 0–9, 10–19, 20–29 ... etc. will not have scope to enter any value between 9 and 10, 19 and 20 and so on.
3. **Open-Ended Class intervals (OECD)**: These are class intervals in which either the lower or upper limits are not distinctively mentioned. For example, class intervals like 85 years and above, less than 100 miles, above 600$, etc.

Steps in creating a grouped frequency distribution table:

1. Locate the largest and smallest values in the data set.
2. Compute the **Range** which is the difference between the maximum value and the minimum value.
3. Divide the data into different class intervals. These intervals must accommodate all the data points. The optimal number of class intervals must be between 5 and 20.
4. Make sure that the minimum value of the first-class interval must be lesser than the minimum value of the entire data set and the maximum value of the last class interval must be higher than the maximum value of the entire data set. This is to accommodate a few more values than the existing values in the given data set.
5. Also, make sure there are no overlapping class intervals of data.

Construction of a frequency distribution

The following steps should be taken into consideration while constructing a frequency table.

- **Class intervals**: The class intervals are obtained by proper division of the range into subranges. The number of class intervals should ordinarily be between 5 and 20. With more than 20 classes, the study may become unnecessarily tedious and with less than 5 classes a great degree of accuracy is lost.
- The class intervals should be of equal width to make calculations easy.

- The class intervals should be such that the structure of the data should not be much altered.
- The size of the class interval should neither be too small nor too large.
- The class intervals should not overlap and should exhaust all the variate values.
- As far as possible open-ended class intervals should be avoided.
- The class limits, mid values and width should be integers or simple fractions to make calculations easy.
- Depending on the variable and degree of accuracy, inclusive or exclusive class intervals may be used. If the values of the variable are discrete, inclusive class intervals will be a better option
- The relative frequency distribution can be obtained by dividing the frequency in each class of the frequency distribution by the total number of observations. A percentage distribution may then be formed by multiplying each relative frequency or proportion by 100.
- The use of relative frequency and percentage frequency distribution becomes essential whenever one bunch of data is being compared with other batches of data especially if the number of observations in each batch differs.

Example 1: Let the test scores of 20 students in a science class be as follows: 23, 26, 11, 18, 09, 21, 23, 30, 22, 11, 21, 20, 11, 13, 23, 11, 29, 25, 26 and 26. The discrete and continuous frequency distribution tables would be (Table 1.1).

Relative Frequency Distribution: Relative frequencies refer to the proportion of frequencies (of discrete or grouped data) to the total frequency. Percentage frequency distribution = Relative frequency * 100.

$$\text{Relative frequency} = \frac{\text{Class frequency}}{\text{Total frequency}}$$

Example 2: A mechanic at his workplace wants to test the battery life of a certain type of battery. He tabulates the relative frequency distribution table to compare the life of batteries under each class interval of its battery life as shown in Table 1.2.

Observation: With this table, we can notice that at least 28% of batteries have a minimum life of 400 minutes.

Example 3: If a surgeon is successful at 6 out of 10 critical bypass operations, then his success rate is 6/10 = 0.6 which is 60%. Tabulating such frequencies along with their respective class intervals is called **relative frequency distribution**, as given above.

III. Cumulative Frequency Distribution:

A cumulative frequency is also called **the running total**, which is the sum of all the frequencies preceding a particular value. A better understanding of the calculation of cumulative frequencies is shown in Table 1.3. This frequency type reveals the total number of frequencies that lie above or below a particular data point.

Table 1.1 Ungrouped and grouped frequency distribution table of marks scored by students

Frequency distribution table		
S. No.	Marks secured	Number of students
1	9	1
2	11	4
3	13	1
4	18	1
5	20	1
6	21	2
7	22	1
8	23	3
9	25	1
10	26	3
11	29	1
12	30	1
Grouped frequency distribution table		
Sl. No.	Marks secured	Number of students
1	0–5	0
2	5–10	1
3	10–15	5
4	15–20	1
5	20–25	7
6	25–30	6
Total		20

Table 1.2 Life of a certain type of a battery in minutes

Battery Life in minutes (X)	Frequency	Relative frequency	Percentage
400–450	15	0.28	27.8
450–500	12	0.22	22.2
500–550	10	0.19	18.5
550–600	9	0.17	16.7
600–650	6	0.11	11.1
650–700	2	0.04	3.7
Total	52		100

Table 1.3 Calculating cumulative frequency distribution

S. No.	Class interval	Frequency	Cumulative frequency
1	0–10	2	2
2	10–20	5	7 (=2 + 5)
3	20–30	10	17 (=7 + 10)
4	30–40	12	29 (=17 + 12)
5	40–50	14	43 (=29 + 14)
Total		43	

Table 1.4 Amount of groceries spent by families in a locality

S. No.	Amount spent on groceries every month	Families	Cumulative frequency
	0–500	0	0
1	500–550	18	18
2	550–600	16	34 (16 + 18)
3	600–650	16	50 (34 + 16)
4	650–700	11	61 (50 + 11)
5	700–750	14	75 (61 + 14)
6	750–800	9	84 (75 + 9)
7	800–850	5	89 (84 + 5)
Total		89	411

Example 4: A grocery store owner in a locality wants to check how many people in the vicinity are willing to spend 650 rupees on groceries every month. He tabulates the following data and calculates the cumulative frequency (Table 1.4 and Fig. 1.5).

Inference: If you plot any cumulative frequencies using a simple line graph as shown in Fig 1.5, we can see an increasing trend line (As the frequencies add up or remain the same.) The cumulative line is usually in "**S**" shape and is also called **ogive**. In this example, we can notice that 50 out of 89 families which is 56.17% (More than 50%) of them in the locality are willing to spend at least 650 rupees on groceries every month. This means if the grocery store provides a wide range of good quality products to its customers, then it has a potential for high sales. With a cumulative frequency graph, we can also understand what is the minimum amount majority of the families are willing to spend on groceries (Table 1.5).

1.2 Scales of Measurement

The way variables are categorized and defined are called scales of measurement. These evaluate certain properties of scales of measurement, which are: **identity**, **magnitude**, **equal intervals** and a **minimum value of zero** (Fig. 1.6).

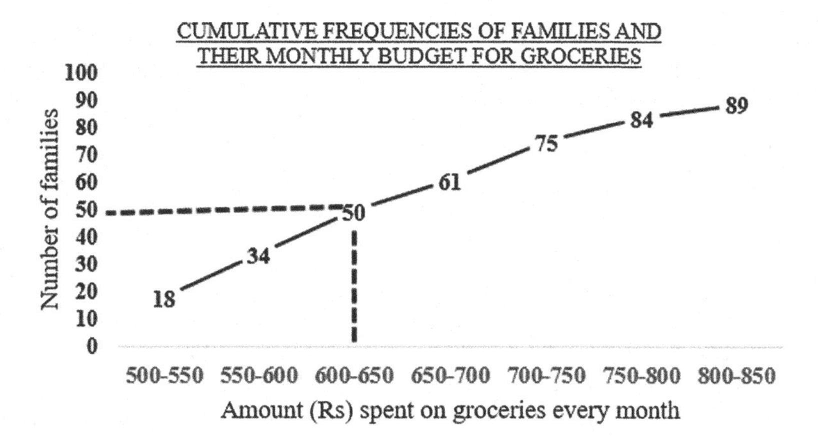

Fig. 1.5 Cumulative frequency line graph for monthly budget for groceries

Table 1.5 Cumulative frequency table for families and their budget for groceries

S. No.	Amount spent on groceries every month	Families	Cumulative frequency
	0–500	0	0
1	500–550	18	18
2	550–600	16	34
3	600–650	16	50
4	650–700	11	61
5	700–750	14	75
6	750–800	9	84
7	800–850	5	89
Total		89	411

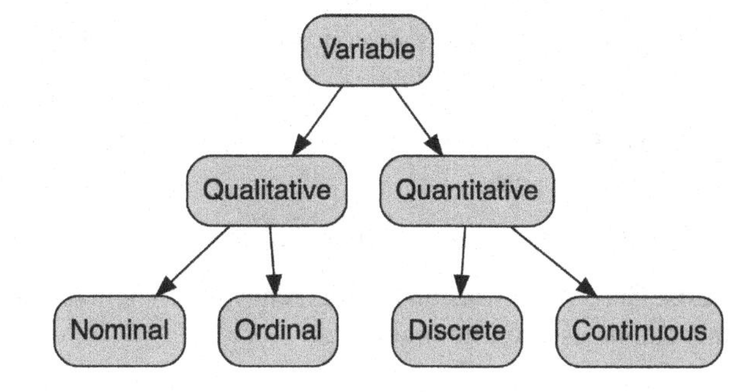

Fig. 1.6 Scales of measurement

The four common scales were developed by psychologist Stanley Stevens in his paper on "**The theory of scales of measurement**" in which he defines four hierarchical scales of measurement namely,

i. Nominal.
ii. Ordinal.
iii. Interval scale.
iv. Ratio scale.

These scales of measurement are in ascending order based on the information that the variable reveals. Though there are many discussions on the pros and cons of this classification, it is widely used in most data analytics. Further, when we use these variables at advanced analytical stages like simulations and forecasting. The measurement scale of a variable will be an important base for the variables under study.

1. **Nominal or classificatory scale**

   - This is the simplest level wherein objects are classified into different categories which are **exclusive as well as exhaustive.** This means that one observation can be put under only one category and there should be enough categories to put in all observations.
   - These categories have to be non-overlapping. Categories can be in the form of numbers too.
   - For example, public transport buses are numbered based on the routes they take in the city and football players of a team are assigned random numbers so that their identity is not revealed.
   - **The codes are meant only for classification and understandability.** When codes such as 0 and 1 are assigned to observations, they are only meant for identification and they **cannot hold mathematical logics** such as 1 is greater than 0 or 0 is lesser than 1, etc.
   - Also, the symbols or codes used can be interchanged without causing any alterations in the data.

Example 5: Hair colour, hobbies, literacy checks, questions with options such as Yes/No/Maybe, etc.

Tools: Mode, proportion and frequency.

Tests: Chi-square tests and contingency coefficients are those statistical operations that can be used for nominal data.

2. **Ordinal scale**

   - This scale is used to indicate the place of a category in an orderly series and they state the position of the characteristic with respect to others.
   - Thus, in an ordinal scale, the **values are hierarchical and state the relative position of a characteristic**.

- Unlike nominal scales, ordinal scales allow comparisons of the characteristics of the variable. Generally, number 1 is assigned to that which has a larger quantity of what is being studied. Just like students are awarded ranks for the marks they score.
- It has to be remembered that **the difference between two levels of an ordinal scale is not the same as the difference between two other levels.**
- For example, if 1 denotes highly satisfied, 2 denotes satisfied and 3 denotes moderately satisfied, then $2 - 1 \neq 3 - 2$.

Example 6: Opinions/perceptions of respondents in surveys are a common example of ordinal scales which are measured on a scale of 5 or 10, grades of students in a class test, military rankings in the army, hierarchical posts in government jobs, etc.

Tools: Median, quartile deviation and Spearman's rank correlation.

Tests: Mann–Whitney U test and Kruskal–Wallis H test.

3. **Interval Scale**

- This is the scale having all properties of nominal as well as ordinal scale. In addition, it has the characteristics of equal intervals.
- This means that the difference between 40 and 50 °F represents the same temperature difference as the difference between 80 and 90 °F. This is because each 10° interval has the same physical meaning.
- This scale has a true zero point even if one of the scaled values is called "***zero***".
- The Fahrenheit temperature scale is an example of this. 0 °F does not mean the absence of temperature. It does not make sense to compute ratios of temperatures. The ratio of 40–20 °F is the same as the ratio of 100–80 °F.

Example 7: Credit scores, pH levels, CGPA and SAT score of students.

Tools: Arithmetic mean (AM), median, standard deviation (SD), and product-moment correlation coefficient.

Tests: Parametric tests like Z- test, t-test, F- test.

4. **Ratio Scale**

- This is the most informative scale.
- It is an interval scale with the additional property that it's zero position indicates the absence of the quantity being measured.
- The ratio scale can be regarded as the successor of all the previously defined scales.
  - Like a nominal scale, it provides a name or label for each object.
  - Like an ordinal scale, the entities are ordered.

  - Like an interval scale, the same difference at two places on the scale has the same meaning.

And in addition, the same ratio at two places on the scale also carries the same meaning.

- The Fahrenheit scale of temperature has an arbitrary zero point and is therefore not a ratio scale. However, zero on the Kelvin scale is an absolute zero value. This makes the Kelvin scale an example of a ratio scale.
- Similarly, money is measured on a ratio scale because, in addition to having the properties of an interval scale, it has a true zero point. If one has zero money, this implies an absence of money. Since money has a true zero point, it is right to say that someone with 50 rupees has twice as much as someone having 25 rupees.

Example 8: Medicine dosage, reaction rates, weights and lengths of aquatic species, etc.

Tools: Geometric mean (GM), coefficient of variation (CV), Karl Pearson's R correlation. ANOVA and which requires knowledge of true scores can be used.

Tests: Both parametric and non-parametric statistical tests can be used (Fig. 1.7).

Contextual Cases

Case I: Consider research in which a questionnaire is administered on their opinion on capital punishment.

Right Tool: Ordinal scale.

The ratings may be on a 5-point scale starting with 1—Highly agree, 2—Agree, 3—Neutral, 4—Disagree and 5—Strongly disagree.

Reason: Ordinal scale is the right choice as there is a hierarchy but the difference between 1 and 2 is not the same as the difference between 3 and 4.

Case II: Consider a saree merchant having a wide variety of cloth materials for sale, like pure silk, cotton, chiffon, georgette, banarasi, etc.

Right Tool: Nominal scale. Where the categories can be marked as

PS—Pure silk, B—Banarasi, C—Chiffon, G—Georgette, CO—Cotton.

Reason: The scale is nominal as these category labels are only used for identification purposes. However, if we arrange these in ascending order of prices, it will be ordinal data.

Case III: Consider a set of data values obtained through a questionnaire along with their levels of measurement.

Data 1: Name of the respondent.

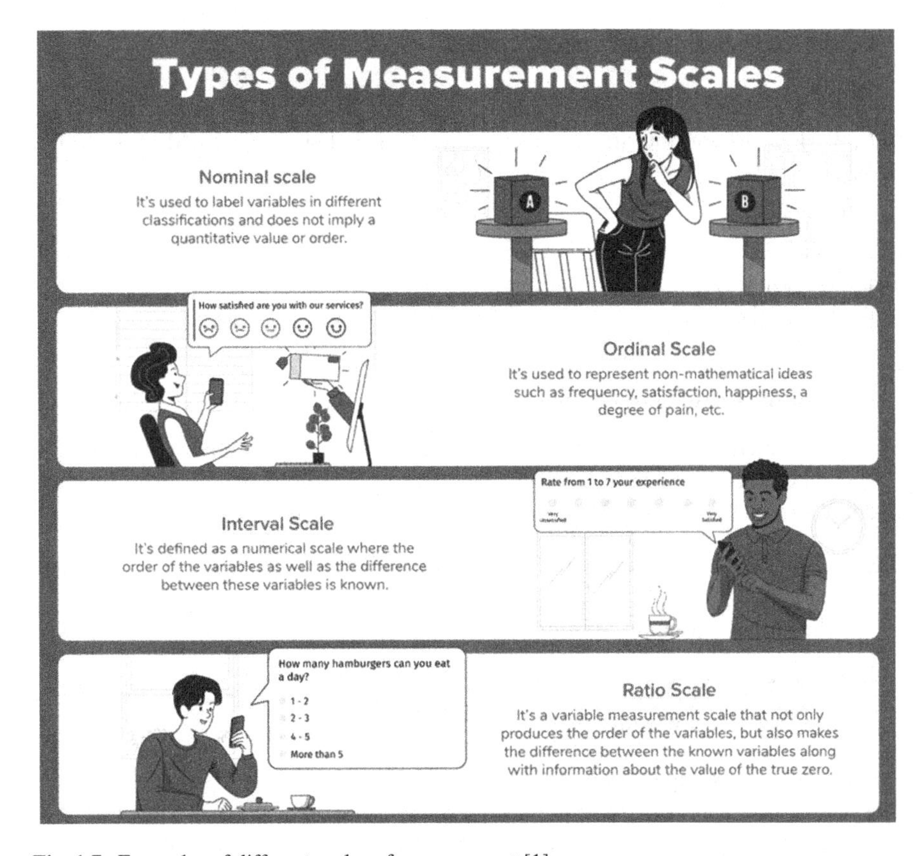

Fig. 1.7 Examples of different scales of measurement [1]

Level of measurement: Nominal.

Why? Since the name does not indicate any order and is just used for identification.

Data 2: Age of respondent.

Why? Since age has a meaningful zero. It is relevant to find the ratio. For example, a 30-year-old person is twice as old as a 15-year-old.

Level of measurement: Ratio.

Data 3: Which is your latest level of academic achievement?

Secondary.

Senior secondary.

Bachelor's degree.

Master degree.

Level of measurement: Ordinal.

Why? These are categories ordered according to hierarchy.

Data 4: How do you rate the new program which is being telecasted at 8 pm on "Crazy TV"?

(i) Very crazy.
(ii) Quite crazy.
(iii) Somewhat crazy.
(iv) Not at all crazy.

Level of measurement: Ordinal.

Why? The choices can be ordered but there are no meaningful numerical differences between them. For instance, "quite crazy" is not double of "somewhat crazy".

Data 6: What is your marital status? 1—Unmarried or 2—Married.

Level of measurement: Nominal.

Why? These numbers 1 and 2 serve as indicators only.

Where can we go wrong? (Tables 1.6 and 1.7)

Example 9: For example, consider a hypothetical study in which 15 girls are asked to choose their favourite colour of hair bands from blue, red, yellow, green and white. An analyst codes the choices as 1, 2, 3, 4 and 5 respectively. This means that if a girl said her favourite colour was Red, then the choice was coded as "2", if she said her favourite colour was White, then the response was coded as 5, and so on. For this, hypothetical data, since each code is a number, we tend to compute the average. In this case, the average is 3.5. The possible interpretation would be that the average favourite colour is a shade between yellow and green. It is very evident that it is a **meaningless conclusion**

Table 1.6 Choice of colours

Choice	Code
Blue	1
Red	2
Green	4
White	5
White	5
Yellow	3
Yellow	3
Green	4
Red	2
White	3

Table 1.7 Summary of scales of measurement

Level of measurement	Some examples	Measurement procedures	Permitted mathematical operations	Permitted statistical operations
Nominal (Categorical)	Marital status, religion, color of dress, type of pen (ink-pen, gel pen etc.) preferred	Classification into non overlapping categories	Counting the number of cases in each category, comparing sizes of each category	Mode, proportion frequency, chi-square, contingency coefficient, tests based on binoansion
Ordinal	Hierarchy in an institution, attitude and opinion scales	Classification into categories as well as ranking those categories with respect to each other	Counting the number of cases in each category, comparing sizes of each category, deciding on "greater than", "equal to", "less than"	Median, quartile deviation, Spearman's correlation, Kendall's correlation, non parametric tests like run test, U test etc
Interval	Fahrenheit, centigrade scales of temperature	All properties of nominal and ordinal as well as a relation between two classes	All the above plus mathematical operations	AM, median, standard deviation, product-moment, correlation, parametric tests like Z, t, F etc
Ratio	Currencies, age, length of buildings, B.M.I of people in a community	All properties of nominal and ordinal as well as a relation between two classes. Also, the ratio between values can be found meaningfully	All the above plus mathematical operations	GM, coefficient of variation, parametric/non parametric tests

as the values mentioned here are just the codes and are used for identification purposes only. These numbers do not hold any mathematical logic.

A note on validity and reliability:

To evaluate and measure an observation, one needs to check if,

Q1: We are measuring what we intend to measure?

—Validity

Q2: Is the measurement process yielding consistent results on multiple repetitions?

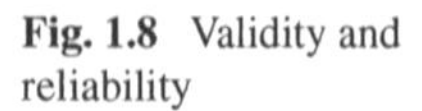

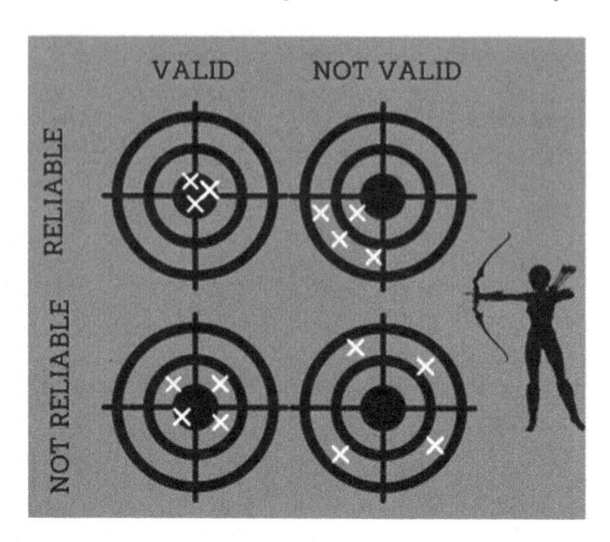

Fig. 1.8 Validity and reliability

—Reliability.

As an analyst, one needs measurement tools/techniques/data models and a process that is both reliable (consistent results) and valid (satisfy the objectives) to fundamentally fulfil the basic expectation of any statistical results delivered (Fig. 1.8).

A reliable measurement is not always valid, whereas a valid measurement is always reliable. This is because results can be reproduced consistently but not necessary that they are correct and precise. Whereas, accurate results are always reliable.

For example, a clinical thermometer records temperatures 2 degrees lesser than normal due to some malfunctioning. It consistently records the incorrect temperature until we get it repaired. These results are consistent but not valid. A pharma company testing the new drug and getting varied responses from the patients is not a reliable drug to be released into the market.

1.3 Representing Data—Graphs and Charts

Displaying data in a meaningful way that is visually appealing through various tools such as graphs, diagrams, maps and charts is called **representation of data**. There are many impressive data graphs, but only a few are purposeful, effective and accurate. As shown in Fig. 1.9, we must always be very cautious in avoiding misinterpretation of data when it is represented visually. Therefore, it is highly essential for us to learn some rules and guidelines for data representations.

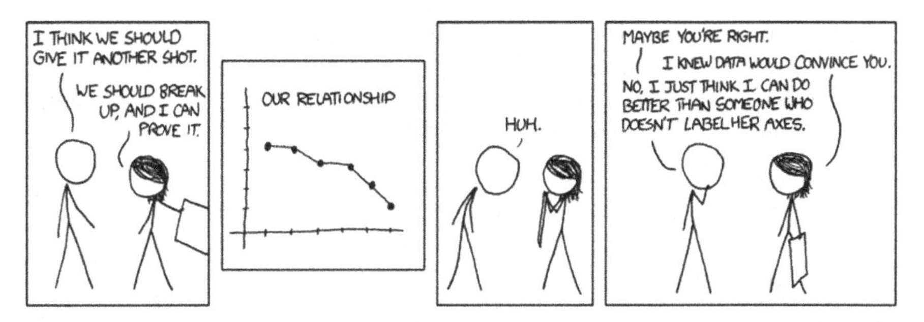

Fig. 1.9 Importance of labelling an axis [2]

Some Common terminologies used in statistical graphs and charts

1. **Title**: A chart title is similar to a story title, which gives the reader a clear understanding of what is depicted in the chart.
2. **Gridlines**: Straight horizontal and vertical lines are used for better interpretation of values on graphs.
3. **Outliers**: A data point is said to be an outlier when it's placed at an abnormal distance from other data points. These are unusual values that stand apart from other data points and sometimes may have a different behaviour too.
4. **Legend**: It contains the required and additional information (if any) about the graph. For example, details regarding the axis, the meaning of different shapes, sizes, patterns and colours (if any) used in the graph, etc. This information by default is shown in a box and placed at the corners of the graph.
5. **Plot area**: The surface area bounded by the axis is called the plot area. It's the region where the graph is plotted.
6. **Dependent and Independent Variables**: Dependent variables are those variables that are affected by the changes made in an independent variable. In any experiment/analysis, the independent variable is the cause and the dependent variable is the effect (Table 1.8).
7. **Axis and Axis Labels**:

Table 1.8 Independent and dependent variables

S. No.	Research topic	Independent variable	Dependent variable
1	Effect of a vegan diet on cholesterol and blood sugar levels	Type of oil used for cooking food	Blood pressure. Sugar levels and cholesterol levels in the body
2	Study on sleeplessness based on the duration of screen time before bed	Screen brightness and screen time before going to bed	Quality and duration of deep sleep
3	Does pea harvest increase with adding Sulphur to the soil?	Amount of sulphur in soil	Quality and quantity of harvest over a period of time

These are lines in different directions that give a framework for the chart. In general cases, X-axis and Z-axis are used for marking qualitative/characteristic features of data. Y-axis is the axis that displays the numerical value. Axis labels are the names of the X, Y and Z-axis that are used in the graph.

- **Horizontal Axis**: X-axis is also called Abscissa.
- **Vertical Axis**: Y-axis is also called Ordinate.

Guidelines to Graphs

Step 1: Determine the dependent and independent variables.

Independent variables are always on the abscissa and dependent variables are always on the vertical axis or the ordinate. Consider an experiment in which diffusion is dependent on the independent variable the temperature. Figure 1.10 depicts the right way of positioning the variables on the axis.

Step 2: Proportion of the axis.

We must try to always keep both axes roughly in a square shape. Exaggerating or cluttering the scales of the axis is generally prohibited. Figure 1.11 depicts the right way of proportioning the axis.

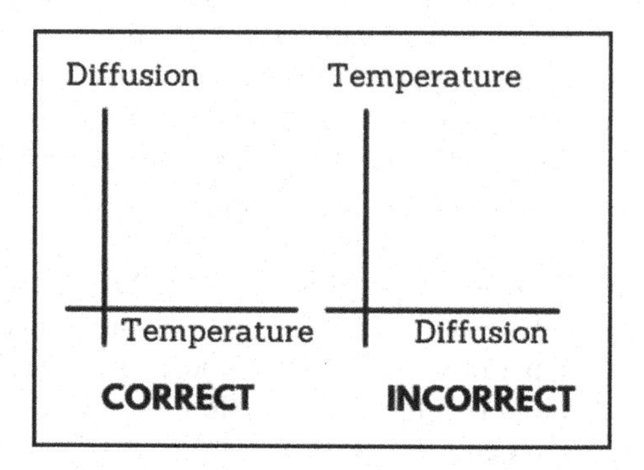

Fig. 1.10 Variables and their axis

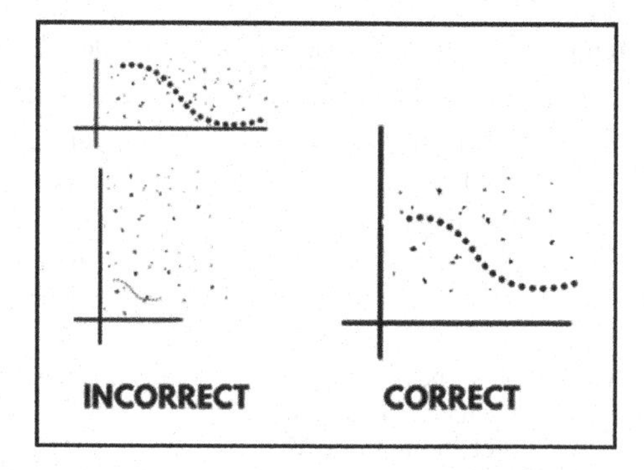

Fig. 1.11 Proportion of axis

Step 3: Units of measurement of the variables to plot.

Mentioning the unit of measurement of variables used in a graph is a vital step that is generally disregarded. Data without units can misguide the analysis. We must always make sure units of measurements are mentioned right next to the label of the axis to avoid any confusion.

Step 4: Plotting data.

1. Calculate the range of the data.
2. Think of a suitable scale for the axis.
3. The labels on the axes should be regularly spaced so that the axis functions as a scale bar for intermediate values.
4. Mark the quantities on both axes and number them at regular intervals.
5. Start from the origin which is 0 in most cases. Sometimes when it's not, the origin must be placed according to the data.
6. Label the graph very clearly. Mention the following.

- Title of the graph.
- An index mentioning all the symbols, patterns and scales used in the graph.
- Axis labels.
- Units of measurements.

Consider Fig. 1.12, which displays all the vital components of a graph. Every graph must have these elements to give the reader a complete understanding without misinterpretation.

Sometimes we may have to plot two data sets on a single axis. In such cases we use symbols like, X, O, ★, etc., or lines such as {....}, {__}, {--}, for differentiation. Sometimes two different dependent variables might have to be plotted together but since the values may be so different that you have to use two different scales. One axis can be placed on each side of the graph as shown below. We must make sure

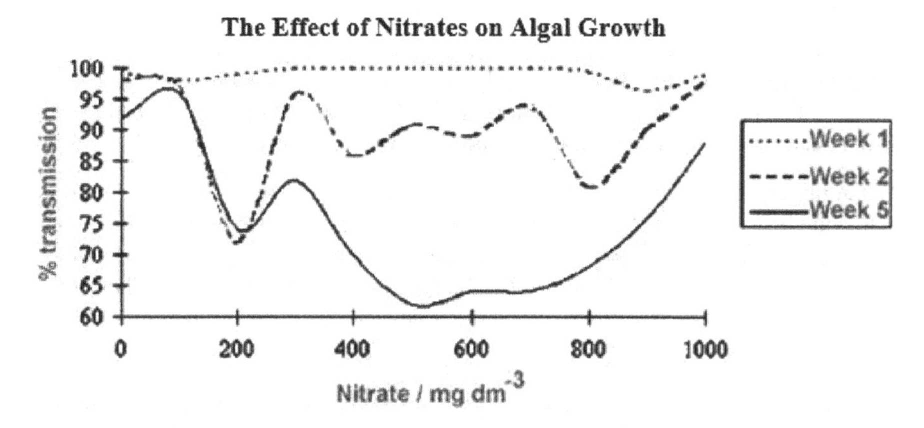

Fig. 1.12 Effect of nitrate on algal growth—graph depicting all the vital components of a graph

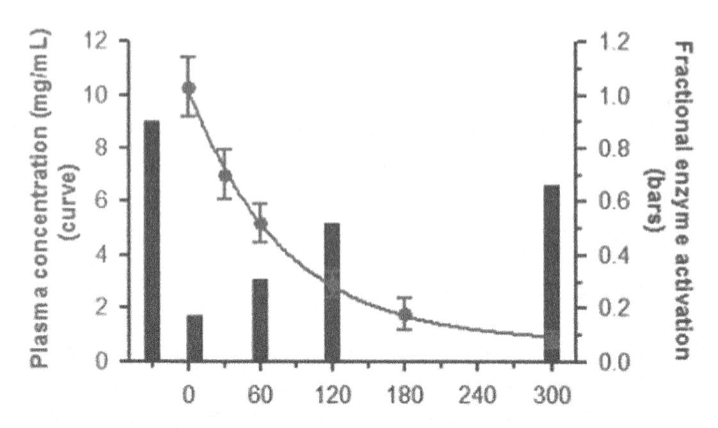

Fig. 1.13 Sample graph with two dependent axes variables on a single axis

that uniform scales must be maintained and all the required information must be given in order to eliminate confusion. Thus, charts must be informative, accurate and impressive in a professional manner. Take a look at Fig. 1.13.

When to use which chart?

It is often very handy and easy to select a common type of visual (like the histogram or scatter plots, pie chart, etc.) from the available templates in the software. As an analyst, one must cautiously choose the type of visual used based on the purpose of representation, which is depicted in Fig. 1.14.

Points to be noted

- Refrain from choosing fancy charts for displaying your data.
- Choose the charts thoughtfully based on the purpose and idea we intend to convey.

Impactful Charts.

"A picture is worth more than a thousand words". However, one should not just use it for the sake of adding colour and variety but has to consider carefully as to which is the appropriate and logical format of data representation. **It should never be necessary to present the same information in two different graphical formats.**

a. Highlighting the important variable under study. This way, the reader's mind is channelized on what is intended to convey. Also, decluttering information adds value to the chart. The above graph represents the non-mortgage debt outstanding. The study is involved in dealing with only student loans and not others. Therefore, depicting all the reasons and highlighting relevant information makes more sense to the readers.

b. Sometimes summary charts are more meaningful than charts with partial information. This may not serve the purpose as all the readers would expect a quick

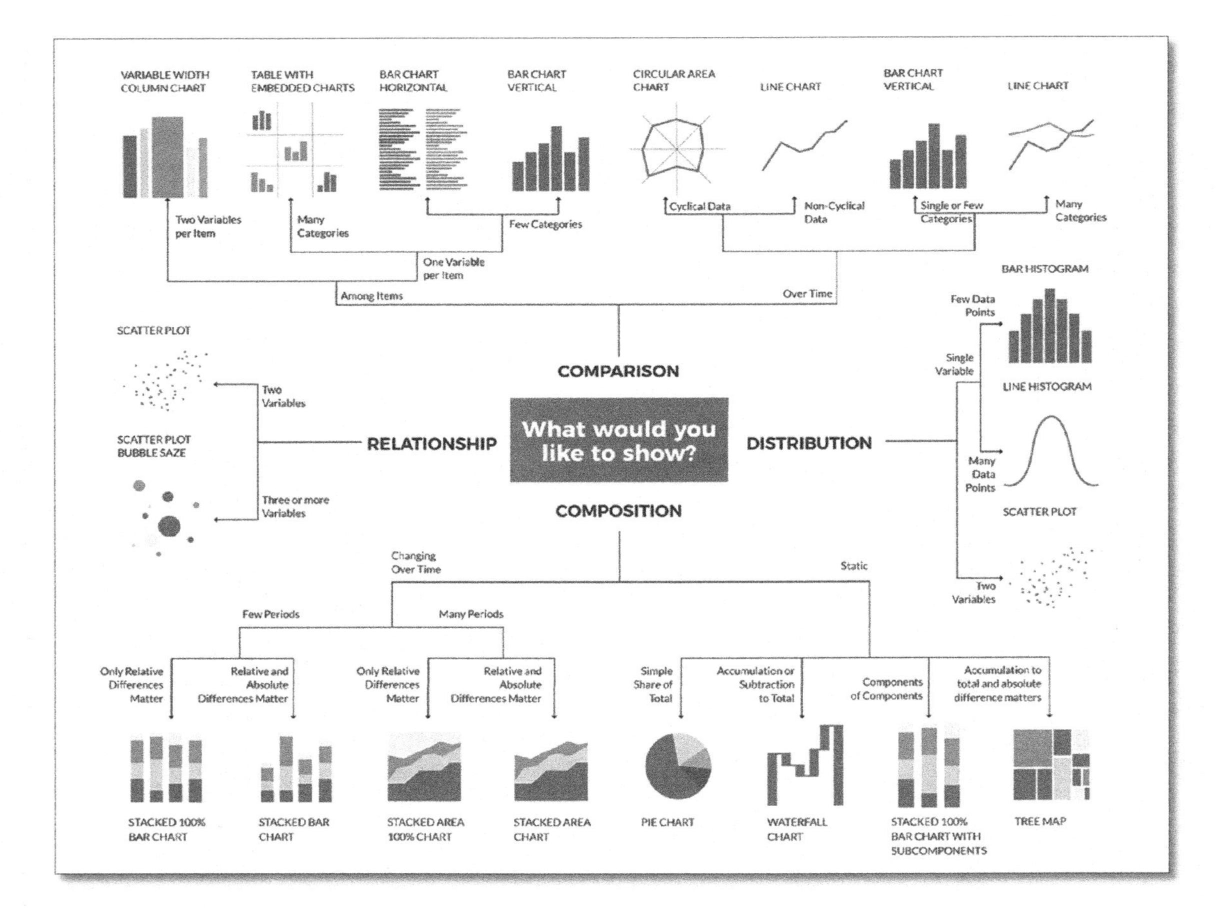

Fig. 1.14 Classification of charts based on its purpose [3]

and clear conclusion when they see the graph. The graphs modelled in this way are effective communicators.

The above chart Fig. 1.15b represents screen time in the USA based on the orientation, landscape or portrait.

(iii) Key data points marked on a chart can also narrate a data story, like how it's depicted in the third chart in Fig. 1.15c. This makes the chart more impactful to the readers as it makes them more curious and allows them to explore the graph elements. The chart shows how the Comcast speeds have reduced the cost for Netflix customers.

Types of Charts and diagrams of data (Fig. 1.16).

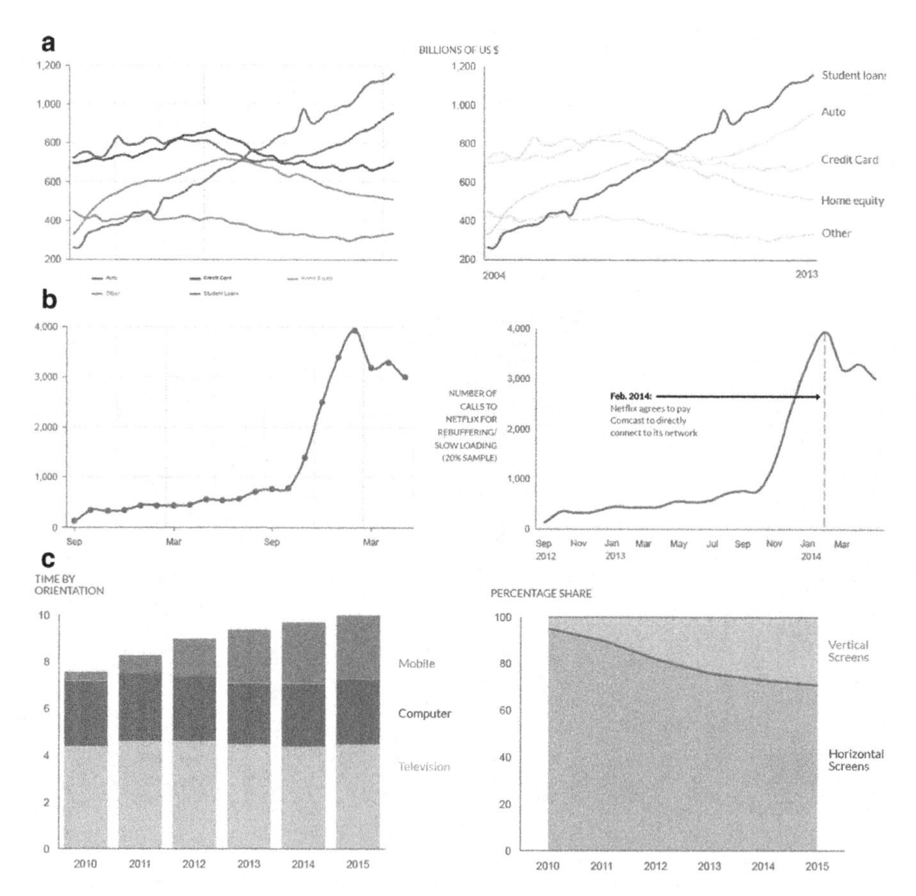

Fig. 1.15 **a** Chart highlighting the important information under study variables on a single axis. **b** Chart that gives out quick information that is required. **c** Chart that that narrates a story

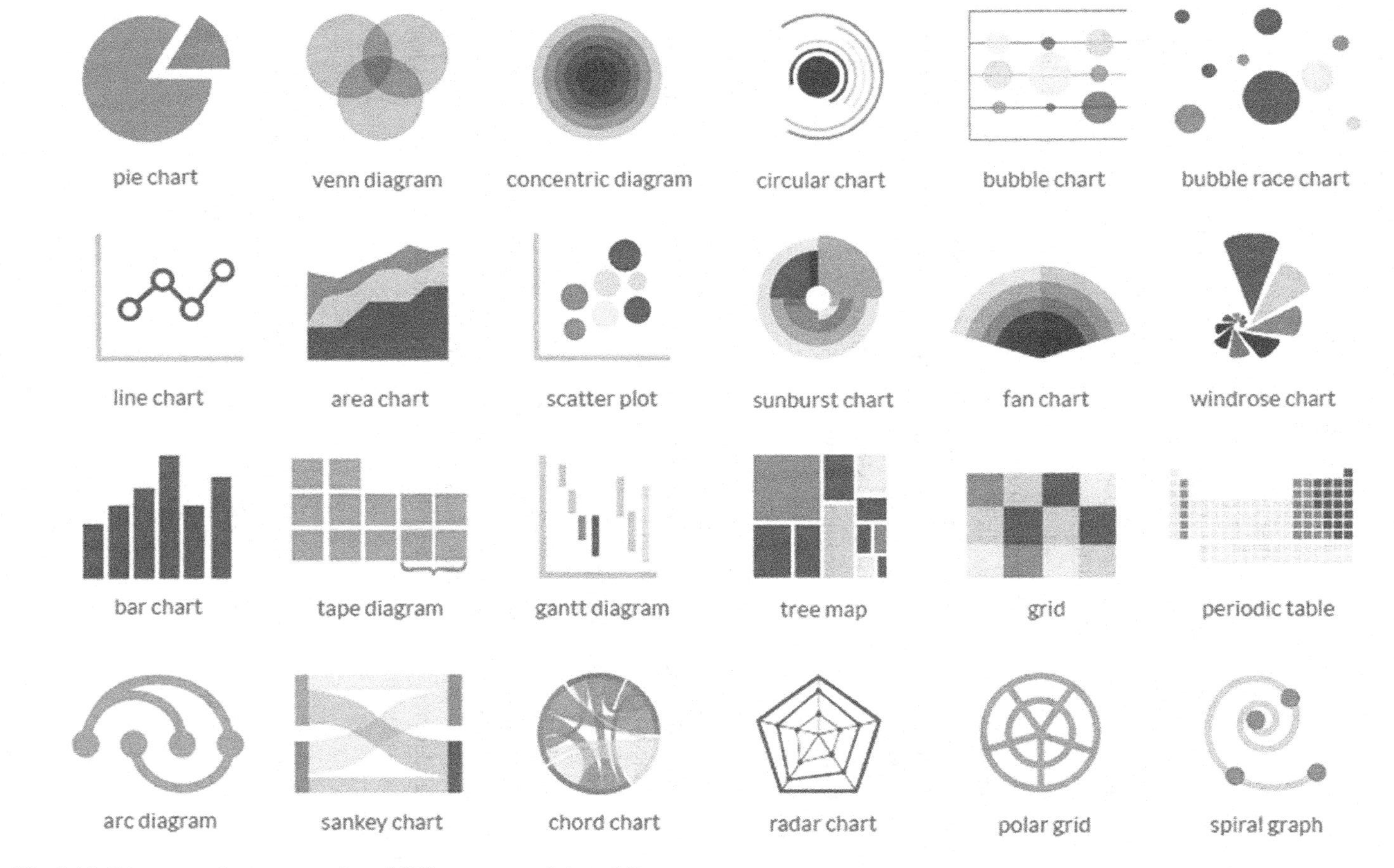

Fig. 1.16 Diagrammatic representation of different types of charts [4]

1.4 Box Plot, Stem and Leaf Diagrams

A box plot or a whisker plot is a very simple yet highly informative visual tool as it reveals the spread and shape of data along with other elementary statistical quantities. Box plots are also used for comparisons between multiple data sets. In this section, let's get a brief idea of what a box plot looks like. In the second chapter, elements and practical usage of box plots are explained in detail. In a box plot, there is a fairly rectangular box representing 50% of data from the midpoint which is called **the median**. The lines next to the boxes are called **whiskers**. They represent the remaining 50% of the data. The endpoints of the whiskers denote the maximum and minimum data points. This difference is the **range** (difference between the highest and lowest values) of the data set. The points beyond this range are called **outliers.** (Which are unusual data points far from the data cluster.) Box plots can be drawn either vertically or horizontally. In Fig. 1.17. The Return on assets (ROA) has been plotted across years for a company. In this picture, we can clearly see the spread of data. **Skewness** refers to the symmetry of the data. Box plots also reveal the symmetry of data as shown in Fig. 1.18. If there are more data points to the right side of the median, there is a **positive skew**. And similarly, if there are more data points to the left of the median, there is a **negative skew** (Fig. 1.19).

Example of box plot:

Consider the total number of tornadoes recorded by the state (including the District of Columbia) from 2000 in the Table 1.9: The total number of recorded tornadoes in 2000, arranged alphabetically by State and including the District of Columbia. Take a look at the box plot representation for the tabulated data (Fig. 1.20).

A note on stem and leaf plots

A stem and leaf plot is another interesting data visualization tool that can be used for both discrete and continuous data. They are also called **diagrammatic representations of frequency distributions**. It is a quick and smart tool for visualization also used for comparisons. Like box plots, they also depict skewness, spread and also central tendencies of data (we will discuss in the second chapter). In stem and leaf plots, always remember to

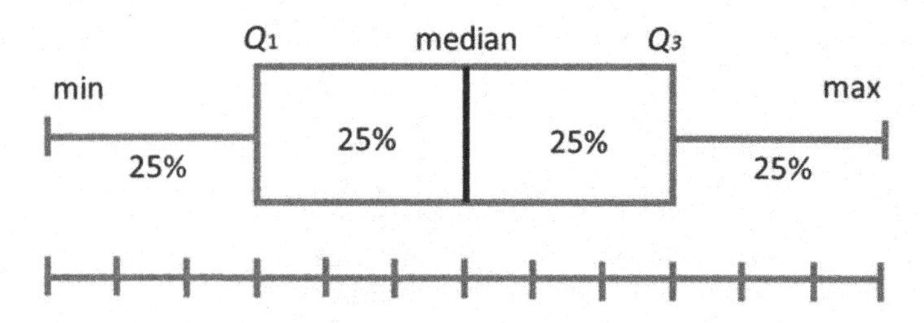

Fig. 1.17 Basics of box plot

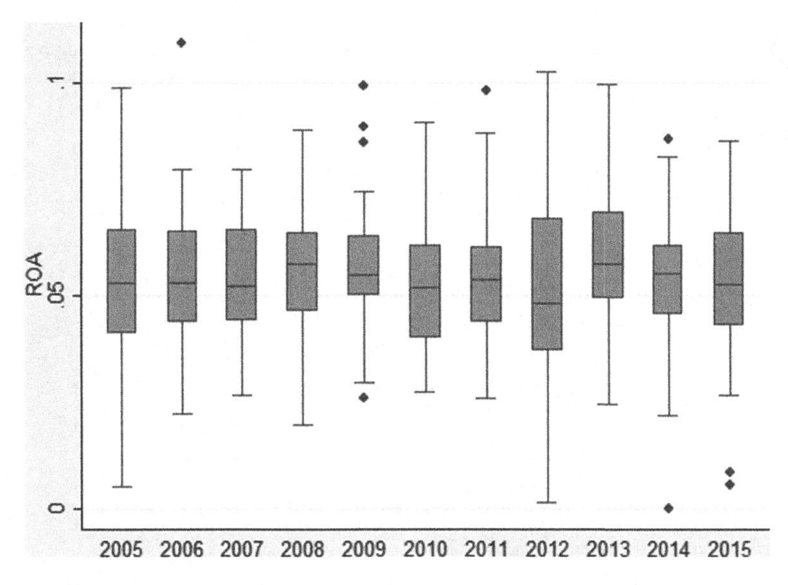

Fig. 1.18 Boxplots for spread of data [5]

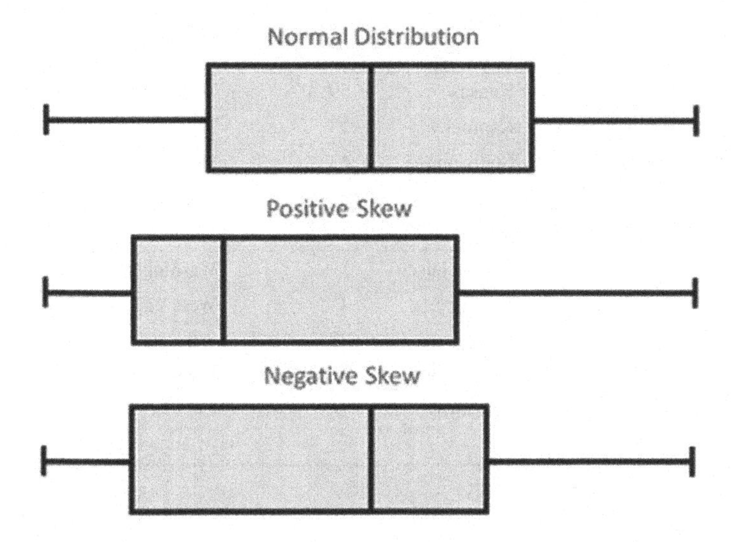

Fig. 1.19 Boxplots for symmetry [6]

- Arrange data (generally in ascending order) before plotting.
- Mention the key of the chart. Which is very important for readers to interpret the data in the plot.

Table 1.9 Statewise list of number of Tornadoes

State	Number of Tornadoes	State	Number of Tornadoes
Alabama	44	Montana	10
Alaska	0	Nebraska	60
Arizona	0	Nevada	2
Arkansas	37	New Hampshire	0
California	9	New Jersey	0
Colorado	60	New Mexico	5
Connecticut	1	New York	5
Delaware	0	North Carolina	23
District of Columbia	0	North Dakota	28
Florida	77	Ohio	25
Georgia	28	Oklahoma	44
Hawaii	0	Oregon	3
Idaho	13	Pennsylvania	5
Illinois	55	Rhode Island	1
Indiana	13	South Carolina	20
Iowa	45	South Dakota	18
Kansas	59	Tennessee	27
Kentucky	23	Texas	147
Louisiana	43	Utah	3
Maine	2	Vermont	0
Maryland	8	Virginia	11
Massachusetts	1	Washington	3
Michigan	4	West Virginia	4
Minnesota	32	Wisconsin	18
Mississippi	27	Wyoming	5
Missouri	28		

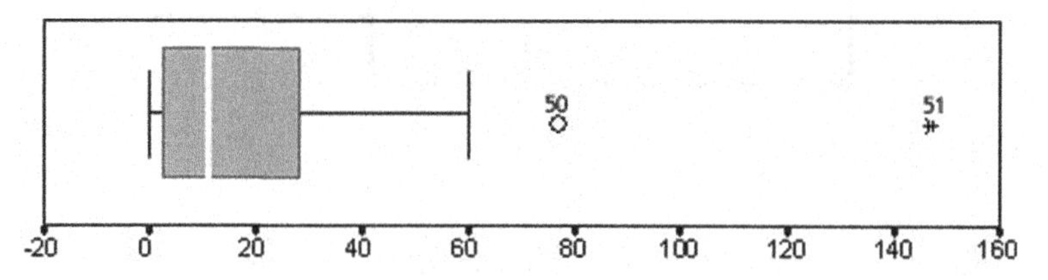

Fig. 1.20 Boxplots for number of Tornadoes [7]

Key in Stem and leaf plot: Consider the number 23. Here the number in the tens place which is 2 is written in the STEM and the number in the unit's place which is 3 is the LEAF. This "|" symbol is used to separate the leaf from the stem elements.

Steps to construct Stem and Leaf plot

- **Step 1:** Arrange the data in ascending order. Take a note of the number of digits.
- **Step 2:** Select a suitable key for the plot, for example, 3|5 = Can be 35 or 3.5 and similarly,15|2 can be 152 or 15.2 as well. Therefore, the key must be mentioned appropriately.
- **Step 3:** Arrange the data in ascending order starting from the lowest to the top and determine the range of the data.
- **Step 4:** List the stems in the stem column.
- **Step 5**: Plot the leaves in the column against the stem from the lowest to the highest horizontally.

Example 10: Consider the scores of 15 actuarial students in a unit test. 10, 13, 21, 27, 33, 34, 35, 37, 40, 40, 41, 33, 48, 26, 39. Tabulate a grouped frequency distribution table with inclusive class intervals, plot a histogram and also show a stem and leaf representation for the same (Fig. 1.21).

Solution:

Stem and Leaf plot for decimal data The weights (kilograms) of 30 students were measured and recorded as follows: 59.2, 61.5, 62.3, 61.4, 60.9, 59.8, 60.5, 59.0, 61.1, 60.7, 61.6, 56.3, 61.9, 65.7, 61.4, 58.9, 59.0, 61.2, 62.1, 61.4, 58.4, 60.8, 60.2, 61.7, 60.0, 59.3, 61.9, 61.7, 58.4, 62.2. Prepare a stem and leaf plot for the data and also give your understanding of the data.

Solution:

In this case, the stems will be the whole number values and the leaves will be the decimal values. The data ranges from 56.3 to 65.7; therefore, we start at 56 and finish at 65.

This stem and leaf plot reveals that the group with the highest number of observations (mode) recorded is in the 61.0 to 61.9 group. In this example, 56.3 and 65.7 can be considered outliers, since these two values are quite different from the other values. Figure 1.22 displays the stem and leaf plot before and after eliminating outliers. However, one must be very cautious while dealing with outliers. Eliminating data has to be done only after carefully considering all the possibilities.

Note on unimodal, bimodal and multimodal data (Fig. 1.23)

Mode refers to the value that is repeated the maximum number of times in the data set. If there is one data point that has maximum frequency it's called **unimodal**. When there are two values with maximum frequencies, then we call it **bimodal** and if there are many more values that are repeated a maximum number of times, it's

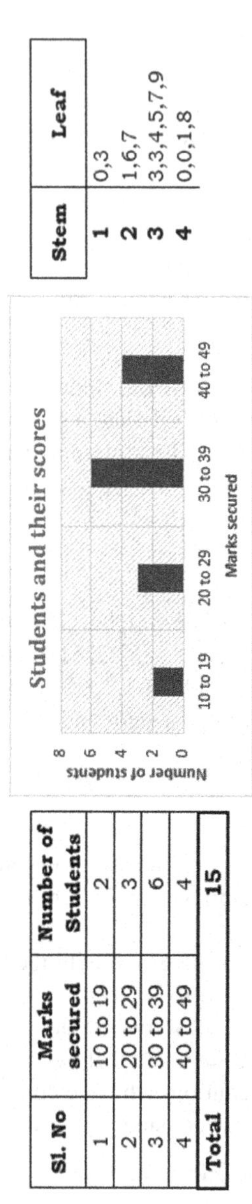

Sl. No	Marks secured	Number of Students
1	10 to 19	2
2	20 to 29	3
3	30 to 39	6
4	40 to 49	4
Total		15

Stem	Leaf
1	0,3
2	1,6,7
3	3,3,4,5,7,9
4	0,0,1,8

Fig. 1.21 Bar chart, stem and leaf diagram for marks scored by actuarial science students

Stem	Leaf
56	3
57	
58	4,4,9
59	0,0,2,3,8
60	0,2,5,7,8,9
61	1,2,4,4,4,5,6,7,7,9,9,
62	1,2,3,7,
63	
64	
65	7

Stem	Leaf
58	4,4,9
59	0,0,2,3,8
60	0,2,5,7,8,9
61	1,2,4,4,4,5,6,7,7,9,9,
62	1,2,3,7,

Fig. 1.22 Stem and leaf plots for weights of students

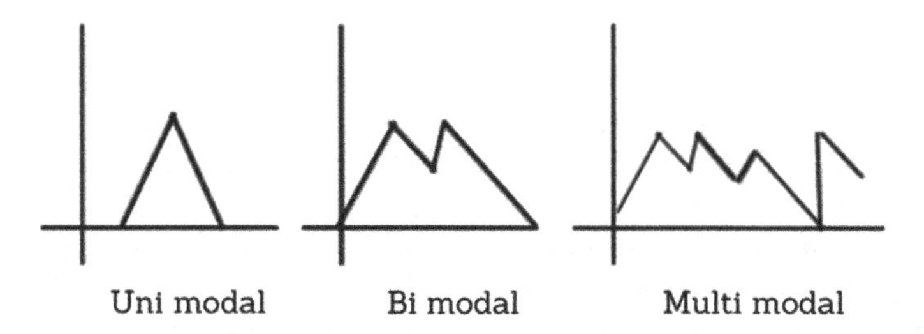

Fig. 1.23 Unimodal, bimodal and multimodal data

called **multimodal data**. The mode of data can be clearly identified in the stem and leaf plot display along with the shape of the distribution of data.

Example 11: These are the scores of 41 students in a math test. (With a best possible score of 70) 31, 49, 19, 62, 50, 24, 45, 23, 51, 32, 48, 55, 60, 40, 35, 54, 26, 57, 37, 43, 65, 50, 55, 18, 53, 41, 50, 34, 67, 56, 44, 4, 54, 57, 39, 52, 45, 35, 51, 63 and 42.

1. Is the variable discrete or continuous? Explain.
2. Calculate the range and prepare an ordered stem and leaf plot for the data and briefly describe what it shows.
3. Are there any outliers? If yes, then which one?
4. Describe the data's distribution with main features such as the number of peaks, symmetry and value at the centre.

Solution:

1. A test score is a discrete variable. For example, it is not possible to have a test score of 35.74542341, as generally, such scores are not awarded.

Stem	Leaf
0	4
1	8 9
2	3 4 6
3	1 2 4 5 5 7 9
4	0 1 2 3 4 5 5 8 9
5	0 0 0 1 1 2 3 4 4 5 5 6 7 7
6	0 2 3 5 7

Stem	Leaf
0	4
1	89
2	346
3	1245579
4	012345589
5	00011234455677
6	02357

Fig. 1.24 Tabulation and visual representation of stem and leaf plot for marks scored by students [8]

2. The lowest value is 4 and the highest is 67. Therefore, the range of the data is 63 and the stem and leaf plot that covers this range of values looks like Fig. 1.24. The key for the data is 24 is represented as **2|4,** where the stem is 2 and the leaf is 4.
3. The value, student securing 4 marks, seems to be like an outlier because there is a big difference between 4 and 18. But this can be for some unknown or uncertain reason. Eliminating such data points is not appropriate without knowing the complete details. 4. There are 41 observations and the distribution of data is slightly concentrated on the right side. The left tail extends farther from the data centre. Therefore, the distribution is skewed 20th and 21st observations to the left or negatively skewed. The distribution has a single peak within 50–59 class interval. Since there are 41 observations, the median value will occur between the which are 45 and 48. So the median is an average of 45 and 48 which is **46.5**.

Example 12: Britney is a swimmer training for a competition. The number of 50-m laps she swam each day for 25 days is as follows:

22, 21, 24, 19, 27, 28, 25, 29, 28, 31, 22, 39, 20, 10, 26, 24, 27, 28, 26, 28, 18, 32, 25, 31, 27.

1. Prepare an ordered stem and leaf plot. Make a brief comment on what the data depicts.
2. Redraw the stem and leaf plot by splitting the stems into five-unit intervals.

Solution:

1. The least and highest value of the data set is 10 and 39; therefore, the range is 29. The plot has stems of 1, 2 and 3. The ordered stem and leaf plot is shown below with key: 1|2 is 12 where 1 is the stem and 2 is the leaf.

Stem	Leaf
1	0, 8, 9
2	0, 1, 2, 2, 4, 4, 5, 5, 6, 6, 7, 7, 7, 8, 8, 8, 8, 9
3	1, 1, 2, 9

Fig. 1.25 Stem and leaf plot for number of 50-m laps by Britney

Fig. 1.26 Revised stem and leaf plot for number of 50-m laps by Britney

Stem	Leaf
$1^{(0)}$	0
$1^{(5)}$	8,9
$2^{(0)}$	0,1,2,2,4,4
$2^{(5)}$	5,5,6,6,7,7,7,8,8,8,8,9
$3^{(0)}$	1,1,2
$3^{(5)}$	9

2. The revised stem and leaf plot shows that Britney usually swims between 25 and 29 laps in training each day (Fig. 1.25).

Note: The stem $1^{(0)}$ means all data between 10 and 14, $1^{(5)}$ means all data between 15 and 19, and so on. The values $1^{(0)}$ $0 = 10$ and $3^{(5)}$ $9 = 39$ could be considered outliers.

Some Case Studies (Fig. 1.26)

Case study: Visual honesty—size matters

It seems that in 1860, there were over eight million milk cows in the USA, and by 1936, there were more than twenty-five million. While picturising the same information, the artist made a big mistake of considering the area for representing an increase in one-dimensional data. This means, that a linear increase in height results in a quadratic increase in area. The two numbers from the data are roughly three times the magnitude of the other but are represented by two cows, one of which is 27 times larger.

Case Study

Have a look at this graph taken from Robert Putnam's "Bowling Alone" Fig. 1.27a. This graph violates the three fundamental rules of data presentation:

1. The chart does not depict meaningful data.
2. The data it depicts is ambiguous.
3. The chart design is seriously inefficient.

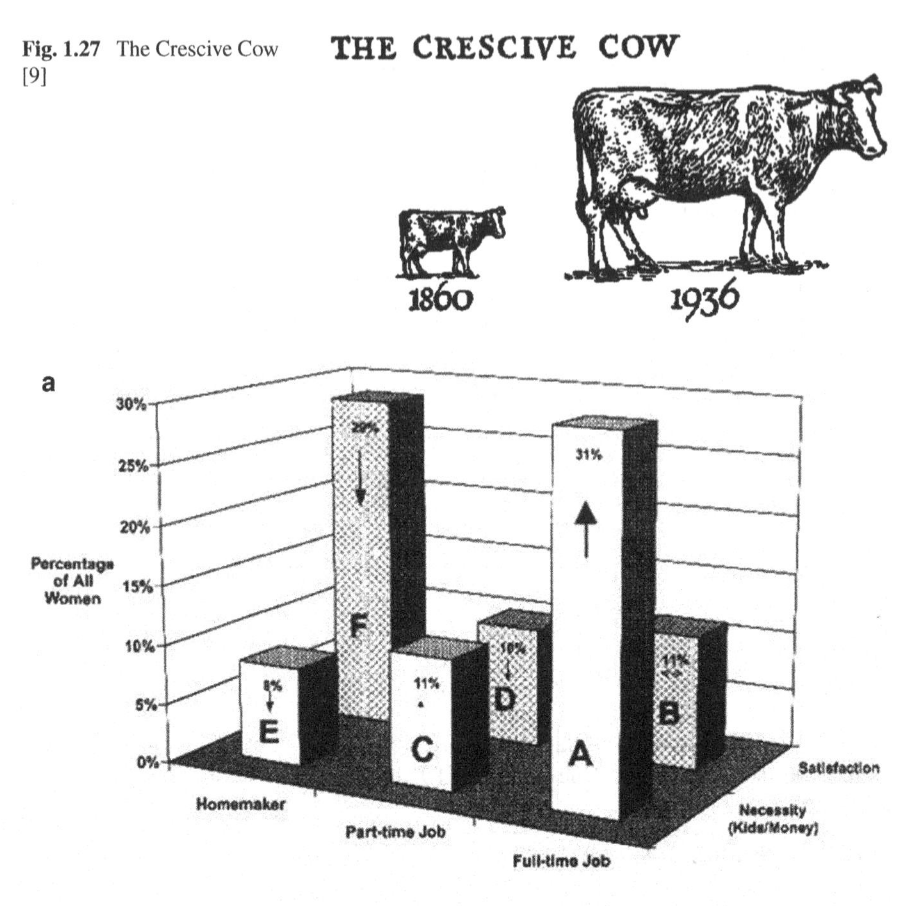

Fig. 1.27 The Crescive Cow [9]

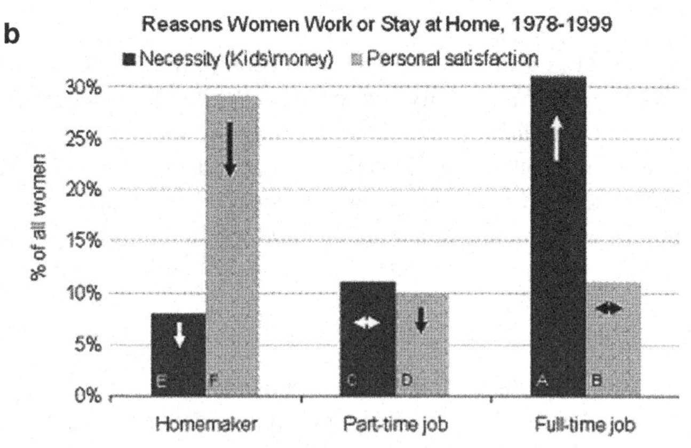

Fig. 1.28 **a** 3D graph depicting women's preference to either work or stay at home. [10] **b** 2D graph depicting women's preference to either work or stay at home

Instead, the graph depicted in Fig. 1.28b is more understandable. Generally, 3D graphs are used for multidimensional data, where the viewpoint gives a better understanding.

For small data sets, 3D graphs are best avoided as they dilute the scale and fabricate the chart elements unnecessarily.

Practice Data Sets—Coding

All the information below and many more links can be found on our GitHub Page:

https://github.com/ElementaryStatisticalMethods/Book1.git

For Python Users:

In Jupyter Notebooks, **Seaborn Matplotlib** is the most popular and easy-to-use library for plotting and also understanding data using descriptive statistics.

Data sets used: IRIS data set

1. https://www.kaggle.com/residentmario/plotting-with-seaborn
2. https://www.kaggle.com/biphili/seaborn-matplotlib-iris-data-visualization-code-1

Other interesting charting tools, techniques and examples with different data sets are provided in these links.

For using the library Matplotlib:

3. https://www.kaggle.com/prashant111/matplotlib-tutorial-for-beginners
4. https://www.kaggle.com/chandraroy/plotting-with-pandas-matplotlib-and-seaborn

All types of graphs and charts are explained with different use cases:

1. https://github.com/rasbt/matplotlib-gallery/tree/master/ipynb
2. https://github.com/VictoriaLynn/plotting-examples
3. https://bookdown.org/dli/rguide/bar-graph.html
4. https://www.oreilly.com/library/view/python-data-science/9781491912126/ch04.html.

For R programming users:

Source of Data: Statista.com.

1. This link provides a brief explanation of all data visualizations with code and examples.

   https://bookdown.org/dli/rguide/bar-graph.html

2. This link provides a brief explanation of descriptive data with use cases of cumulative, relative frequency distribution tables with examples.

   https://data-flair.training/blogs/descriptive-statistics-in-r/

References and Research Articles

1. **Scales of measurement**: https://www.questionpro.com/blog/nominal-ordinal-interval-ratio/
2. **Importance of labelling axis, data Comic**: https://xkcd.com/833/
3. **When to use which chart?** https://education.microsoft.com/en-us/learningPath/d3d1bee5/course/0a60eeb6/1
4. **Types of charts and diagrams**—https://blog.adioma.com/how-to-think-visually-using-visual-analogies-infographic/
5. **Boxplots**—https://www.researchgate.net/figure/7Roe-over-time-box-plot_fig5_318611990
6. https://www.simplypsychology.org/boxplots.html
7. **Boxplots for tornadoes**
8. Adapted from: National Oceanic and Atmospheric Administration's National Climatic Data Centre http://nationalatlas.gov/articles/mapping/a_statistics.html#three
9. **Stem and leaf examples**—https://www150.statcan.gc.ca/n1/edu/power-pouvoir/ch8/5214816-eng.htm
10. http://nationalatlas.gov/articles/mapping/a_statistics.html
11. **The Crescive Cow**—https://sites.google.com/site/dlampetest/blog/thecrescivecow

Chapter 2
Measures of Central Tendencies

WHAT

Measures of Central Tendency,
Dispersion, Skewness and Kurtosis.

WHY

- To condense a data set into a representative value.
- To visualise the spread and variability of data.

HOW

- Analysing data by breaking down data into smaller divisions.
- By exploring the data using various statistical tools.
- Through mapping data points and visualising the shape and spread of data.

WHEN

- The concepts defined in this chapter are useful in comprehending data at the second phase.

S. Prasad, *Elementary Statistical Methods*,
https://doi.org/10.1007/978-981-19-0596-4_2

CHAPTER 2

- Measures of Central tendencies.
- Partition values.
- Dispersion of data.
- Lorenz curve and its interpretations.

WITH MR.STAT

WITH MISS TICS

- Dealing with data sets in both R and Python for comprehensive learning of concepts.
- Dispersion of data used to predict fraud in financial services.
- Exploring the US College scorecard data to understand the central tendencies and partition of data.

Introduction: When we analyse data, we would wish to have a single value that represents or summarizes all the data points in a meaningful way. With this value, we can also understand the tendency of a data to cluster around a typical value. This is precisely what a central tendency means. It is a statistical tool that defines/describes the entire data set in one single value which will also help to describe the data set as well as compare different data sets. The tendency of data values to cluster around a middle value is known as "Central Tendency".

The different measures of central tendency.

1. Mean—Arithmetic, geometric and harmonic mean.
2. Mode.
3. Partition Values—Quartiles, Deciles and Percentiles, where the second quartile is the median.

The researcher must understand the meaning and implication of these values thoroughly before selecting them. The best way to comprehend and compare these measures is by understanding their applications.

2.1 A Note on Summation

This Greek symbol "$\sum$" Summation is a very commonly used symbol for summing values. Huge arithmetic expressions can be simplified into a single line using this summation symbol. In this book, we will come across a lot of summations.

The function inside the summation is iterated starting with the index value given below the summation sign till the condition on top of the summation is satisfied. The final result is the total of all intermediate values we obtain after every iteration. As shown in Fig. 2.1, we start with $k = 1$ and add up all the intermediate results of $k * f(x)$ until $k = N$.

Example 1 Consider the example shown

In Fig. 2.2, we start with $n = 1$ and end when $n = 4$

First iteration: $\boldsymbol{n} = 1$, $[(2 * 1) - 1] = \mathbf{1}$

Second iteration: $\boldsymbol{n} = 2$, $[(2 * 2) - 1] = \mathbf{3}$

Third iteration: $\boldsymbol{n} = 3$, $[(2 * 3) - 1] = \mathbf{5}$

Fourth iteration: $\boldsymbol{n} = 4$, $[(2 * 4) - 1] = \mathbf{7}$

We end the iterations here as $\boldsymbol{n}$ has reached 4. Summation, adds all these values and returns the output: $1 + 3 + 5 + 7 = \mathbf{16}$.

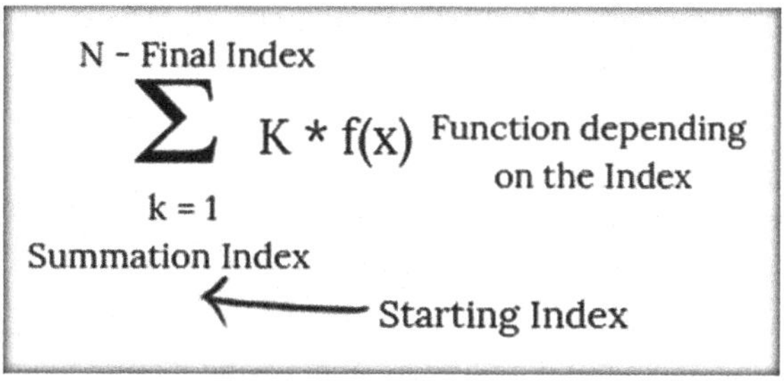

Fig. 2.1 Parts of summation

$$\sum_{n=1}^{n=4} 2n - 1$$

Fig. 2.2 Example of summation

Arithmetic Mean

Arithmetic mean is the most popular and extensively used measure of central tendency. For raw or ungrouped data, the arithmetic mean of a set of observations is their sum divided by the number of observations.

Therefore, the arithmetic mean of ***n*** observations is given by:

$$\overline{x} = \frac{1}{n}(x_1 + x_1 + \cdots + x_n) = \frac{1}{n}\sum_{i=1}^{n} x_i$$

For grouped data, the arithmetic mean is defined as,

$$\overline{x} = \frac{f_1x_1 + f_2x_2 + \cdots + f_nx_n}{f_1 + f_2 + \cdots + f_n} = \frac{\sum_{i=1}^{n} f_ix_i}{\sum_{i=1}^{n} f_i}$$

where f is the frequency and x is the midpoint of the corresponding class intervals.

Note: Arithmetic mean is vulnerable to the influence of outliers

See Figs. 2.3 and 2.4.

Arithmetic mean is sensitive to outliers because even a single outlier can dilute or drag the mean away from its actual value. This results in inaccurate inferences and unrealistic predictions. So, it is always the primary duty of an analyst to eliminate the outliers before computing the arithmetic mean. (Outliers must be carefully eliminated only after considering the reason behind the existence of such values.)

Example 2 The market average of wages is 3,000 per worker. The management of factory ABC claims that they are paying higher than the market average to its staff. Suppose the wages (In thousands) of ten randomly selected workers are as follows, is the claim justified?

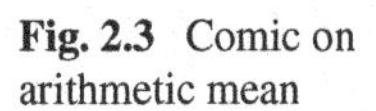
Fig. 2.3 Comic on arithmetic mean

Fig. 2.4 Effect of outliers on arithmetic mean

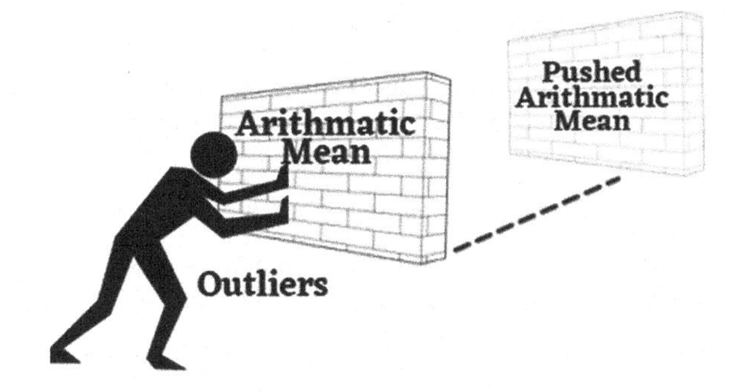

Solution

The mean wages of all 10 staff members is Rs. 33,300 which means on average each staff member gets 3300 rupees which is above the market average. But the salary of the fourth staff member which is 75,000 rupees is probably the value of the C.E.O's salary. This looks like an outlier. If this value is eliminated, the average salary of the remaining 9 staff members is 28,667 which means that each staff member gets an average salary of Rs. 2866, which is below the market average (Rs. 3000). This way management's claim of paying higher wages than the market average to its staff members is not justified.

Example 3 Consider a student's marks in 5 subjects. Based on this data, which stream of education would probably suit him? Engineering or Bachelor of Arts? (Tables 2.1 and 2.2).

Observe that the mean marks are found to be 74.2 (which we obtain by adding all marks and dividing by the total number of values, i.e. 10. If we keenly observe the scores, we can notice that though the average marks of the student are good. This is mainly due to his performance in languages rather than core subjects like mathematics. So, though he may be a polyglot, fields of engineering and core science may be challenging for him when compared to courses on philosophy, language or history. So, opting for a Bachelor of Arts would suit him more than pursuing engineering.

Table 2.1 Wages of 10 random workers at a factory

Staff	1	2	3	4	5	6	7	8	9	10
Salary (thousands)	20	33	36	75	32	30	25	27	22	33

Table 2.2 Student's score in 5 subjects

Subject	French	Kannada	Mathematics	Sanskrit	Hindi
Scores	77	82	40	92	80

Combined Arithmetic Mean

This concept of mean is used when we need a single score or combined mean for two or more sets of data. The formula is given by:

$\overline{X_c} = \frac{n_1\overline{X_1}+n_2\overline{X_2}}{n_1+n_2}$ X_1 and X_2 represent two different data sets with n_1 and n_2 terms in each. $\overline{X_c}$ represents combined mean of both the data sets.

Caution: If we need to find the combined average of two different data sets, say heights in inches and heights in centimetres, we cannot directly use the formula.

For example, Let the average age of men in a firm be 35 years and that of women be 33 years. It is not right to just calculate AM = (35 + 33)/2 = **34 years**. It is always wise to ask for additional information such as the number of men and women in each group.

Case Study: Consider a survey where we need to figure out approximately what age group of people are developing heart diseases. The large population is divided into urban, semi-urban and rural populations. Will arithmetic mean be a suitable tool?

Right tool: The answer is NO, it is not just AM but combined AM. We could also think of weighted AM by taking the number of people in different populations. Using combined AM, we can know the approximate age at which people develop heart disease, by taking into account the number of people in each group into consideration.

Reason: Combined AM is an appropriate measure as the mean ages can be calculated separately and they can be combined using the formula for AM. This is a property of AM that combined mean can be found for different groups. This is not possible for other averages like median and mode. This algebraic property of AM helps in a large number of cases, especially while dealing with different groups like population divided into various strata, people divided according to their age and so on.

Weighted Arithmetic Mean

In certain situations, we have to give more importance to a few entities and less importance to others, for example, in an engineering entrance exam, an institute may give higher weightage for Mathematics than marks in English; i.e., every entity cannot be treated as same. When different weights are associated with the entities, the arithmetic mean is defined as:

$$W_X = \frac{w_1x_1 + w_2x_2 + w_3x_3 + \cdots + w_nx_n}{w_1 + w_2 + w_3 + \cdots + w_n}$$

If there are **n** categories of weights, where **w** is the weight and $x_1, x_2, \ldots, x_n$ are different categories to which $w_1, w_2, w_3 \ldots w_n$ are respective weights associated.

W_X: Total weighted average.
w_nx_n: nth weight * nth category of X.

Table 2.3 Appraisal scores for employee 1 and employee 2

Assessment	Importance	Employee 1 score	Employee 2 score
Skill	0.45	15	19
Knowledge	0.3	11	16
Interpersonal relationships	0.15	17	13
Punctuality	0.05	18	10
Grooming	0.05	13	16
Total		74	74

Example 4 A class of 30 students took a science test. 20 students had an average (arithmetic mean) score of 80. The other students had an average score of 60. What is the average score of the whole class?

Solution
The right tool to be used is weighted AM.

$$\text{Total class average} = \frac{(20 * 80) + (10 * 60)}{30} = \frac{2200}{30} = 73.33\text{marks}$$

Case Study 1
Two employees are assessed on different parameters for their final appraisal and their relative importance is shown below. The scores range from 1 to 20 (20 being the highest). Which measure is most suited to decide his final appraisal score? (Table 2.3)

Right tool: Weighted AM

Reason: Since the average score is important to determine his final score, AM is the right choice. Each category of his assessment is important and is not equal. This requires weighted AM.

Calculation:
See Table 2.4.

Inference: Thus, it is obvious that employee 2 scores much higher than employee 1. The total scores of both employees are **74**. But, when we consider the weighted average of employees 1 and 2 we get 14.15 and 16.6, respectively. We find that employee 2 is more skilled at work which is the main criteria for appraisal, others such as grooming and punctuality are important but are not given much importance

Table 2.4 Calculation for appraisal employee 1 and employee 2

Assessment	Importance	Employee 1 score	Employee 2 score	Weights * Score	
				Employee 1	Employee 2
Skill	0.45	15	19	6.75	8.55
Knowledge	0.3	11	16	3.3	4.8
Interpersonal relationships	0.15	17	13	2.55	1.95
Punctuality	0.05	18	10	0.9	0.5
Grooming	0.05	13	16	0.65	0.8
Total		74	74	14.15	16.6

during appraisal. So, this is the fact we must understand about associating weights in different scenarios.

Case Study 2

At a certain hospital, nurses are given performance evaluations to determine eligibility for merit pay raises. Nurses are rated on a scale of 1–10 (10 being the highest) for several activities: appearance, record-keeping, promptness, and bedside manners with patients. The average is determined by giving weights:

- 5 for promptness.
- 6 for record-keeping.
- 3 for appearance and,
- 7 for bedside manners with patients.

Out of 10, what is the average rating for a nurse with ratings of 7 for promptness, 8 for record-keeping, 6 for appearance and 9 for bedside manner?

Right tool: Weighted AM

Reason: Since the average score is important to determine her final score, AM is the right choice. Each category of his assessment is important and is not equal. This requires weighted AM.

Calculation:

See Table 2.5.

The weighted arithmetic mean is **7.8095**. Thus, weighted AM is very important to evaluate the correct value of AM. Without considering weights, it is not correct to evaluate the measure of central tendency.

Median

Table 2.5 Calculation for appraisal using weights

Evaluation factors	Weights (W)	Scores of nurse (S)	W * S
Promptness	5	7	35
Record-keeping	6	8	48
Appearance	3	6	18
Bedside manners with patients	7	9	63
Total	21	30	164

Median is one of the positional averages. It is the value which divides the data into two equal parts. This means, it is the value that exceeds and is exceeded by the same number of observations, the value such that the number of observations above it is equal to the number of observations below it. The first and foremost step to be followed before calculating the median is to arrange in ascending (it works for descending order too) of magnitude. Median is the best average used when data is skewed. In the case of **raw data**, if the total entities (n), then we first arrange the data and calculate median in the following manner.

1. **Odd Number**: The middle value is the median.
   Example: Consider this data set (arranged in ascending order): 15, 22, 34, 46, 55, 65, 78.
   Solution: The median is the middle term which is **46**. There are three numbers arranged in order on both sides of 46.
2. **Even Number**: The median is the average of two middle terms.
   Example: Consider this data set (arranged in ascending order): 15, 22, 34, 46, 55, 65, 78 and 99.
   Solution: The median is the average of 2 terms which is 46 and 55.
   So, the median is: (46 +55)/2 = **50.5**.

Median for Discrete Frequency Distributions

If **N** is the total number of values and f **is** the respective frequency, the median value will be: $\left(\frac{N+1}{2}\right)^{\text{th}}$ term We total $N + 1$ frequencies and obtain the middle value, for this purpose we need to calculate cumulative frequency at each step, and the value of frequency corresponding to the less than cumulative frequency just greater than half the total is the median.

Step 1: Find "less than" cumulative frequency (L.C.F)

Step2: Find $(N + 1)/2$, where $N = N = \sum_{i=1}^{n} f_i$

Step3: Locate this value, if it exists, or the next higher value in L.C.F column

Step4: The corresponding value of the variable is the median.

Example 5 A farmer wants to calculate the median for his produce. He has made a note on the days (X) and the yield in kilograms (f).

Solution

See Table 2.6.

$(N + 1)/2 = 122/2 = 61$st term. The cumulative frequency which is slightly more than 61 is 65 and the corresponding X value is 21.

Therefore, the median days for the yield of a farmer are 21.

Remember that the median is derived from the Value of variable (X) column and not from any other column.

Median for Continuous Frequency Distributions

Ensure that the class intervals must be exclusive type; however, they need not be of equal size.

The class corresponding to the cumulative frequency equal to or just greater than $2(N/4) = N/2$ is called the median class and the value of the median is obtained by the following formula.

$$\text{Median} = l + \frac{h}{f} * \left[\frac{N}{2} - c\right]$$

l The lower limit of the median class.
f The frequency of the median class.
h The magnitude (size) of the median class (upper limit-lower limit).
c The cumulative frequency of the class preceding the median class.

Step 1: Calculate "less than" cumulative frequency for the entire data.

Table 2.6 Calculation of median for a yield

Days (X)	Yield (f)	Less than cumulative frequency (L.C.F)
10	4	4
11	6	10
15	10	20
16	11	31
20	16	47
21	**18**	**65**
25	26	91
26	30	121
Total	121	389

Step 2: Locate the median class by finding out the value that is equal to or, the next higher value than the $(N/2)^{th}$ value in the cumulative frequency column.
Step 3: Plug in the values into the formula, given above and hence figure out the median value.

Example: Calculate the median for the following data.

Class interval	Frequency
0–10	8
10–20	10
20–30	23
30–40	52
40–50	7
50–60	5
Total	105

See Table 2.7.
Median class is recognized as 30–40 since $(N/2)^{th}$ value is (105/2) = **52.5**.

$$\text{Median} = l + \frac{h}{f} * \left[\frac{N}{2} - c\right]$$
$$= 30 + (10/52) * (52.5 - 41) = 30 + 2.11 = \mathbf{32.11}$$

Usage: Median is the preferred average when

(1) Dealing with qualitative data, arranging the data in order of magnitude makes sense. For example, to find the average intelligence or average honesty levels among criminals and so on.
(2) A distribution is skewed.
(3) There is a reason to believe that the distribution might be skewed.
(4) There are a small number of subjects under the study.
(5) Problems concerning wages, distribution of wealth, etc.

Table 2.7 Problem and solution for median of grouped frequency distribution

Mid value	Class interval	Frequency	Cumulative frequency
5	0–10	8	8
15	10–20	10	18
25	20–30	23	41
35	**30–40**	**52**	**93**
45	40–50	7	100
55	50–60	5	105
Total		105	365

Case Study 1: Here is a famous case of median versus mean
In early April 2005, there was a considerable debate in the UK media about whether "average incomes" had gone up or down. The Institute for Fiscal Studies produced a report in which they stated that the mean "real household income" fell by **0.2% over 2003–04** against the previous year. The shocking thing is that when this was reported in the media, some commentators were glorifying this apparent reduction of average incomes as an opportunity to criticize the government. (Gordon Brown, who was the chancellor at the time, was very frustrated trying to explain that the median is the measure you use for things like income because the distribution is skewed.)

Right tool: Median

Reason: If the distribution is not symmetrical, it is always ideal to use median rather than mean. If more people earn low salaries and we use mean, the analysis will be biased. Observe in the picture that the median is in the middle of mode and mean in both positively and negatively skewed distributions [2].

Case Study 2
For the patients who experience a heart attack caused by a complete blockage of blood supply to the heart (called ST-segment elevation myocardial infarction or STEMI), quick response is critical. These patients should be treated with artery-opening procedures soon after they arrive at an emergency room. As a result of the complete blockage of the artery, virtually all the heart muscles being supplied by the affected artery starts to die if the patient is not treated quickly.

In the heart attack care community, the mantra is "**time is muscle**", referring to the urgent need to restore the flow of oxygen-rich blood to preserve the heart's function. A measurement that is also important is the door-to-balloon time (D2B): The time between going through the door of an emergency room to the time a balloon is used in opening the blocked artery (the procedure is called angioplasty) (Fig. 2.5).

The American Heart Association (AHA) and American College of Cardiology (ACC) recommend a door-to-balloon time of no more than 90 min for STEMI.

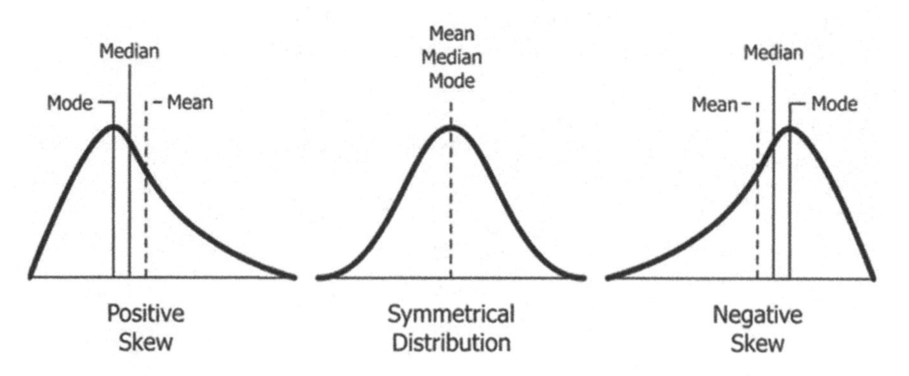

Fig. 2.5 Median of skewed data

Thanks to a nationwide concerted effort, the median door-to-balloon time dropped from 96 min in 2005 to 64 min in 2010. This is a success story in healthcare quality improvement.

For angioplasty to be performed promptly for a STEMI heart attack, many things have to fall into place. The staff to perform the procedure needs to arrive on time (preferably within 20–30 min). Should angioplasty be determined to be necessary; the catheterization lab needs to be activated (preferably with a single phone call). For the heart attack care unit to be efficient and effective there has to be strong administrative support and all the personnel have to work with a team mentality. A mechanism for providing real-time feedback, a way for the staff to share experience and to find out what went right and what went wrong can help the staff make continual improvement.

A study conducted in 2006 set out to identify the problems and issues in the low-performing hospitals and the best practices in the high-performing hospitals in treating STEMI heart attacks. The strategies described in the previous paragraph are some of the best practices identified in this study. The widespread adoption of these best practices, as a result of a nationwide concerted effort, has led to the marked improvement in STEMI cardiac across the USA.

The most important improvement is on the key quality measure of door-to-balloon time. According to a new study, door-to-balloon time declined from a median of 96 min in the year ending 31 December 2005, to a median of 64 min in the 3 quarters ending 30 September 2010. This study represents a 5-year follow-up. A 4-year follow-up published one year earlier also presented a similarly excellent report card.

According to the same new study, the percentage of patients treated under the 90-min guideline has also increased substantially. The percentage of patients treated with angioplasty within 90 min increased from 44.12% in 2005 to 91.4% in 2010. Furthermore, the percentage of patients treated within 75 min increased from 27.3% in 2005 to 70.4% in 2010. The investigators reported that the declines in median times were greatest among groups that initially had the highest median times:

- Patients older than 75 years of age (median decline 38 min),
- Women—35 min, and
- Blacks—42 min.

Right tool: Median

Reason: In this case, the median is the right measure to be used as it provides the correct average time for treating a patient. Mean will be inaccurate as even a delay of one patient will decrease the mean treatment time. Mode also is not suitable as the most repeated value can only give an idea of the treatment times, but will not denote the time where treatment matters.

Case Study 1

Consider a customer satisfaction survey with the following data. Sample sizes are 9 and suppose they rate their overall satisfaction scores on a scale of 1–10. Let the scores be 1, 2, 3, 4, 4, 7, 9, 10, and 10 when arranged in ascending order.

Right tool: Median

Reason: Mean of scores = 5.55. Suppose the general assumption is that customers above a score of 4 remain loyal, the company has a reason to feel happy as the mean is much higher.

The median of this group is a 4. This means that 5 out of 9 were unhappy. Thus, the company cannot be proud of its customer services since (5/9) = 55.5% of customers are unhappy.

Case Study 2

In his post Mind the Gap, Steve Murchie asserted that the VC industry has effectively stopped investing in seed-stage ($500 K and less) and start up-stage ($2 M and less) opportunities. The following data was given (Table 2.8).

The mean would give a wrong picture as there are outliers in this data, so the median will be a better choice. The median of % of Total $M is the average of the two middle terms, 29.48 and 18.49 which is **23.98**. Now, let us consider the same data with additional information (Table 2.9).

Observe that the fifth column "Average/Deal $M" is free from outliers, and thus, the arithmetic mean tool can be used. Though the data is still skewed, it is much

Table 2.8 Dollars funded for a start-up at different stages

Stage	Total $M	% of total
Later stage	1522	37.49
Expansion	1630	29.48
Early stage	1121	18.49
Start-up/seed	490	14.54

Table 2.9 Calculation of median value of funding

Stage	Total $M	% of total	No. of deals	Average/Deal $M
Later stage	1522	37.49	148	10.28
Expansion	1630	29.48	175	9.31
Early stage	1121	18.49	189	5.93
Start-up/seed	490	14.54	88	5.56

more balanced than the previous one. The mean of Average/Deal $M is 31.08/ 4 = **7.77**.

Case Study 3

The "**Maui News**" gave the following costs in dollars/ day for a random sample of condominiums located throughout the island of Maui, an island in Hawaii.

95, 70, 59, 66, 356, 75, 520, 75, 45, 295, 66, 55, 280, 49, 49, 175, 255, 90, 66 and 120

Calculating the measures as per definition, we observe that Mean is $143.05, median is $75 and mode is $66.

Right tool: Median

Reason: If mean is used, it will show a steep price which might put off tourists. The mode does not give the right picture as only 4 out of 10 condominiums have this rate. However, the median gives the best average here as the tourist can decide whether he/she can afford to rent one.

Case Study 4

[3] The impact factors provided in Journal Citation Reports (JCRs) have been used as a tool for librarians, authors and administrators to assess the relative importance of journals. Eugene Garfield and Irving Sher developed the Impact Factor in the 1960s to select journals for the Science Citation Index. Data was collected from Thomson Reuters JCR Science Edition for the year 2007. For each of the 172 fields listed, the field's median Impact Factor, median Article Influence Score, highest Impact Factor and highest Article Influence Score were recorded. Median values were collected to represent typical journals in different fields.

Right tool: Median

Reason: The median rather than the mean was used because impact factors tend to be skewed making the numeric average less representative of a typical journal in a field.

Note:

- A commonly committed error is that for data that in a normal distribution, the mean should be used and for data that is not normally distributed, the median should be used. But this is not valid. Mean and the median will be close if the data is normally distributed, else, both the mean and the median will convey useful information.
- Consider a variable X that takes the value 1 for males and 0 for females. Here, we can calculate the mean as well as the median. The proportion of males/females in the group can be known through the mean, whereas through the median, we come to know which group contained more than 50% of the people.
- Mean from ordered categorical variables can be more useful than the median, the ordered categories have meaningful scores. For example, a conference might

be rated as 1 being poor to 5 being excellent. Here, the mean is better than the median.

Mode

Is the typical value that has maximum concentration of values around it? The only situation in which the mode may be preferred over the other two measures of central tendency is when describing discrete categorical data. The mode is preferred in this situation because the greatest frequency of responses is important for describing categorical/nominal data. Some data sets can have,

- One value repeated the maximum number of times, this is called **Unimodal Data**.
- Two values repeated the maximum number of times, this is called **Bimodal Data**.
- More than two values are repeated the maximum number of times, this is called **Multimodal data**.

In the case of **Raw (ungrouped)data**, Mode is the value that is repeated the most number of times in the data.

Example 6 Consider choosing a team from a group of students with the first names Lucky, Dhanush, Rahul, Nandu, Lina, Tony, Lucky, Nandu and Lucky.

Solution
The mode is Lucky as this name occurs most frequently. There is no meaning in finding any other measure of central tendency in this case.

Example 7 Consider the following items sold in an Ice cream parlour: Chocolate, Butterscotch, Butterscotch, Strawberry, Vanilla, Chocolate, Butterscotch, Vanilla, Vanilla, Chocolate, Strawberry, Strawberry, Butterscotch, Chocolate, Chocolate.

Solution
The mode for the above data is chocolate which is the most preferred item by customers in the store. Also, this indicates that the store should stock more chocolates.

Example 8 Consider a study involving mortality rates in a population. The generally used measure is the AM or the life expectancy. However, the mode is also a useful measure to indicate the age at which maximum deaths occur.

Example 9 Consider the distribution of retirement age. 52, 53, 53, 55, 55, 57, 58, 58, 59, 59, 60.

Solution
The median is 57 years and the mean is 56.27 years, whereas the mode is 55 years.

Example 10 An Internet cafe conducts a survey and arrives at the following data for a particular month (Table 2.10).

Table 2.10 Table displaying usage of internet by customers in a cafe

Sl. No.	Usage	No. of customers
1	E-mail	475
2	Chatting	420
3	Browsing	355
4	Downloading	325
5	E-shopping	240
6	E-commerce	195
7	Entertainment	159
8	Adult sites	225
9	Astrology and matrimonial	170
10	Others	125

Right Tool: Mode

Reason: Customer preference is the important information as the cafe can plan for expansion and take up quality improvement projects based on it. Computing the mean and median makes no sense. The mode can easily pinpoint the maximum usage and help the cafe owners.

Example 11 A restaurant conducts a study on the type of places people like to visit and the following options are given along with other questions.

1. Restaurant.
2. Fast Food (multinational).
3. Fast Food- Indian style.
4. Food Court.
5. Local small joints.
6. Pizza outlets.
7. Home delivery.

Right Tool: Mode

Reason: If the above information has the number of people who eat out as well as the total number in the locality, the mean can give an idea about the number of people who like eating out. Median will not serve any purpose and mode can exactly indicate the most preferred way of eating out.

Mode for Discrete Data

1. Find the data with the highest frequency.
2. The value of the data corresponding to this highest frequency is the mode.

Example 12 The following table represents the number of times that 100 randomly selected employees ate at the office cafeteria during a randomly chosen month (Table 2.11).

Table 2.11 Table displaying number of times employees ate at cafeteria

Number of times (X)	2	3	4	5	6	7	8
Number of students (f)	13	41	52	39	10	8	12

Table 2.12 Median marks of students

Marks	0–10	10–20	20–30	30–40	40–50
No. of student(s)	16	10	10	25	14

We observe that the highest frequency is 52. The corresponding value of X is 4. Therefore Mode = 5, the employees are most likely to eat in the cafeteria 5 times a month.

Mode for Continuous Data

- Check whether the class intervals are of continuous type.
- The class intervals must be of equal size.

Identify the maximum frequency in the table and the associated class interval is called the **modal class**. This class interval holds the maximum value in its range. Plugin the values into the formula given to calculate the mode for continuous data.

$$\text{Mode} = L + \frac{f_1 - f_0}{2f_1 - f_0 - f_2} * i$$

f_0 Frequency of pre-modal class.
f_1 Frequency of modal class.
f_2 Frequency of class succeeding the modal class.
L Lower limit of the modal class.
i Size of Class intervals.

Example 13 Consider measuring marks of 75 students arranged in class intervals. Calculate the mode for the following data (Table 2.12).

Solution
The class interval highlighted is the modal class since it has the highest frequency of 25.

Marks	0–10	10–20	20–30	**30–40**	40–50
No. of student(s)	16	10	10	**25**	14

$$\text{Mode} = L + \left\{ \frac{f_1 - f_0}{2f_1 - f_0 - f_2} \right\} * i$$

Table 2.13 Different measures and their usage

Measurement scale	Suitable measures
Nominal or categorical	Mode
Ordinal	Median
Interval	If data is symmetrical, use mean If data is skewed, use median
Ratio	If data is symmetrical, use mean If data is skewed, use median

$$\text{Mode} = 30 + \left(\frac{25 - 10}{(2 * 25) - 10 - 14}\right) * 10 = 30 + 5.76 = \mathbf{35.76}$$

When a data set is bimodal or multimodal, we say that mode is ill-defined. In such cases, we use an alternative formula to calculate mode.

Mode = 3 Median − 2 Mean

Which tool to use when?

See Table 2.13.

Geometric Mean (GM):

Definition
The geometric mean of ***n*** observations is defined as the *n*th root of the product of the ***n*** observations.

Raw data: Let be the n observations. Then the geometric mean is defined as,

$$\text{GM} = \sqrt[n]{x_1 * x_2 * x_3 * \ldots * x_n}$$

Adding log on both sides

$$\text{Log GM} = \frac{1}{n}\log(x_1 * x_2 * x_3 \ldots * x_n)$$

$$\text{Log } GM = \frac{1}{n}(\log x_1 + \log x_2 + \log x_3 \ldots + \log x_n)$$

$$\text{GM} = \text{antilog}\left(\frac{\sum \log(x_i)}{n}\right)$$

Example 14 Find the GM for the data set—15, 48, 75, 62 and 85.

$$\sqrt[5]{15 * 48 * 75 * 62 * 85} = 49.072$$

Applications of Geometric Mean

- Whenever the data deals with rates, ratios and percentages, we use GM. In such cases, AM is not a suitable measure. For example, the ratio of false and true entities in the set can be written as true/false or false/true. In such cases, GM is more suitable as it does not depend on the way the ratio is expressed, which is not the case with AM. Thus, GM is used for calculating index numbers.
- Other examples include the depreciation of machines, growth of bacteria, cell divisions, variations in population sizes, etc. It can also be thought of as a special case where log X is used instead of X when the data is highly skewed. This transformation results in a fairly symmetric distribution of values.
- GM should be used whenever the data is interrelated. Economists use GM rather than AM to calculate the average annual return on investment.

Note: Geometric mean cannot be calculated for negative values.

Grouped data: Consider the discrete frequency distribution, $i = 1, 2, 3\ldots k$ where f_i is the frequency of the variable x_{i} where $\sum_{i=1}^{k} f_i$.

The geometric mean is given by:

$$\text{GM} = \sqrt[n]{f_1x_1 * f_2x_2 * f_3x_3 * \ldots * f_kx_k}$$

Adding log on both sides

$$\text{Log GM} = \frac{1}{n}\log(f_1x_1 * f_1x_1 * \ldots * f_kx_k)$$

$$\text{Log GM}\frac{1}{n}(f_1 \log x_1 + f_2 \log x_2 + \ldots . + (f_k \log x_k +)$$

$$\text{GM} = \text{antilog}\left(\frac{\sum f_i \log(x_i)}{n}\right)$$

Example 15 Consider discussing returns on investment or interest rates. Arithmetic means are not suitable in this case (Table 2.14).

Solution

The geometric mean of the ROI is,

$$\sqrt[5]{0.25 * 0.18 * 0.2 * 0.08 * 0.15} = 0.1609$$

Table 2.14 Table displaying ROI across years

Year	2010	2011	2012	2013	2014
ROI (%)	25	18	20	8	15

The geometric mean is a more valid measure. This is because values are in percentages and so AM will not give a true picture of change. GM = 16.09%, which means that the ROI has risen by an average of 16.09% from 2010 to 2014.

Geometric Mean for Grouped Data

Let $x_1, x_2, \ldots, x_k$ be the class mark/midpoint of k class intervals. The respective frequencies would be denoted as $f_1, f_2, \ldots, f_k$. where $\sum_{i=1}^{k} f_i$.

The geometric mean is given by:

$$\text{GM} = \sqrt[n]{f_1x_1 * f_2x_2 * f_3x_3 * \ldots * f_kx_k}$$

Adding log on both sides

$$\text{Log GM} = \frac{1}{n}\log(f_1x_1 * f_1x_1 * \ldots * f_kx_k)$$

$$\text{Log GM} = \frac{1}{n}(f_1 \log x_1 + f_2 \log x_2 + \cdots + (f_k \log x_k +)$$

$$\text{GM} = \text{antilog}\left(\frac{\sum f_i \log(x_i)}{n}\right)$$

Example 16 The table below provides the annual returns (In % of age) on various investments of insurance companies. Calculate the average returns using the geometric mean.

Returns (% in age)	Frequency
15–25	18
25–35	16
35–50	24
50–60	32
60–65	10
Total	100

Solution See Table 2.15.

$$\text{GM} = \text{antilog}\left(\frac{\sum f_i \log(x_i)}{n}\right)$$

$$\text{GM} = \text{antilog}\left(\frac{159.78}{100}\right)$$

$$\text{GM} = \text{antilog}\ (1.5978) = 30.69$$

Table 2.15 Table for geometric mean of returns on investment (% of age)

Returns (% in age)	Frequency (f)	Mid-point (X)	Log (X)	f * Log (X)
15–25	18	20	1.30	23.42
25–35	16	30	1.48	23.63
35–50	24	42.5	1.63	39.08
50–60	32	55	1.74	55.69
60–65	10	62.5	1.80	17.96
Total	100		7.95	159.78

Some Applications of GM in Real Life

1. It is used in photography and videography to get a balance between aspect ratios.
2. Geometric mean-based citation counts are considered to be most suitable for comparison.
3. It has many applications in the field of medicine.
4. In the field of finance, geometric mean is considered to be suitable to estimate expected returns when the time horizon is long as compared to shorter time horizons, where arithmetic mean is more appropriate.
5. Used for the development of various indices and standards.

Harmonic Mean

The harmonic mean is defined as the reciprocal of the average of the reciprocal of observations.

Raw data: Let $x_1, x_2, x_3, \ldots, x_n$ be n observations. The harmonic mean is given by

$$\textbf{Harmonic mean(H)}\frac{1}{\frac{1}{n}\sum_{i=1}^{n}\frac{1}{x_i}} = \frac{n}{\sum_{i=1}^{n}\frac{1}{x_i}}$$

Harmonic Mean for grouped data:

Let X, be the value of data, in case of discrete and the midpoint of the class interval, in case of continuous data. Consider the frequency distribution $x_i | f_i$, $i = 1, 2, 3, \ldots, n$. The harmonic mean is given by

$$\mathbf{H} = \frac{\mathbf{N}}{\sum_{i=1}^{n}\frac{f_i}{x_i}} \text{ where, } \mathbf{N} = \sum_{i=1}^{n} f_i$$

Situations, where harmonic mean, are extensively used:

- While calculating P/E ratios of any financial institution.
- For calculating the Fibonacci series.
- For calculating averages such as rates and speed.

- Calculating fluctuations in census population data.
- A situation where a large population with a majority of the values is distributed uniformly but there are a few outliers with significantly higher values. In such cases, HM is the right measure to use.

Example 17 Consider the data points 5, 7, 9, 12, 15 and 21.

Solution

The harmonic mean for these values is given as

$$\frac{6}{\left(\frac{1}{5}+\frac{1}{7}+\frac{1}{9}+\frac{1}{12}+\frac{1}{15}+\frac{1}{21}\right)}=\frac{6}{0.666}=9$$

Example 18 A person drives for a half-hour at 120 km/h, and then for a half-hour at 60 km/h. Find his average speed.

Note: The average speed for the trip is the arithmetic mean of the two speeds or 90 km/h.

But, if he drove for 75 km at 110 km/h, and then drove for another 40 km at 75 km/h, the average speed for the trip will be the harmonic mean.

Solution

Time = Distance/Speed. Let's calculate it based on the distance covered.

So, the X_1 = 110/75 = 1.46 and X_2 = 40/75 = 0.53. So, the harmonic mean is **0.78**.

Example 19 A farmer used low nitrogen fertilizer for his pea plantations and made a note of how many pea pods sprouted. He wants to calculate the mean for his data (Table 2.16).

Solution

Table 2.16 Harmonic mean for pea production per plant in Farmer's farm

Number of pea pods per plant (X)	Number of plants (f)
15	5
18	4
19	6
20	6
21	8
24	7
26	9
143	45

Number of pea pods per plant (X)	Number of plants (f)	$1/X$	$f(1/X)$
15	5	0.07	0.33
18	4	0.06	0.22
19	6	0.05	0.32
20	6	0.05	0.3
21	8	0.05	0.38
24	7	0.04	0.29
26	9	0.04	0.35
143	45	0.35	2.19

The harmonic mean for average pea production $= \frac{\sum f * \frac{1}{x}}{N} = \frac{45}{2.19} = 20.54$ which is approximately 21. **Weighted harmonic mean** for X_1, X_2,..., X_n values with corresponding weights w_1, w_2, ..., w_n is given as,

$$\text{Weighted harmonic mean} = \frac{\sum_{i=1}^{n} w_i}{\sum_{i=1}^{n} \frac{w_i}{x_i}}$$

Example 20 A common measure of the relative value of a company's stock is the price-to-earnings ratio. A high P/E ratio indicates that a company's stock is overvalued, or the investors are expecting high growth rates shortly. A relatively low PE indicates that the company's stock is undervalued or the investors are expecting poor/low growth rates shortly. Consider the following companies, for which we want to determine an aggregate PE:

Company	Market capitalization	PE ratio
A	72,42,300	19.5
B	22,30,000	22.1
C	1,95,000	27.25
D	18,250	55

Right tool: Weighted Harmonic Mean

Reason: We are averaging ratios and not numbers (Table 2.17).

If the companies have similar market values, then the average PE would be the harmonic mean. But, if the company's market values are not similar, we need to use weighted HM. If company A comprises a big chunk of the total market value of the four companies, its PE should be given greater weight. Thus, the weighted harmonic mean will be the right tool.

Solution

Weighter harmonic mean $\frac{\sum_{i=1}^{n} w_i}{\sum_{i=1}^{n} \frac{w_i}{x_i}} = \frac{10}{0.3246} = 30.80.$

Table 2.17 Weighted harmonic mean of market capitalisation and P/E ratio of companies

Company	Market capitalisation	PE ratio X_i	Weights W_i	$\frac{W_i}{X_i}$
A	72,42,300	19.5	1	0.0513
B	22,30,000	22.1	2	0.0905
C	1,95,000	27.25	3	0.1101
D	18,250	55	4	0.0727
Total			10	0.3246

Relationship between Harmonic, Arithmetic and Geometric Mean [5]

See Fig. 2.6.

All the 3 averages are very popular in data analytics. A good analyst chooses the appropriate statistical tool at the right time. Here is a key to remember when to use, which average.

- Use the arithmetic mean: When the values have the same units.
- Use the geometric mean: When the values have differing units.
- Use the harmonic mean: When values are in terms of rates.

Note: Harmonic mean and geometric mean uses only positive nonzero values.

The relationship between AM, GM and HM is given as:

$$\text{AM} \geq \text{HM} \geq \text{GM}$$

$$\text{GM} = \sqrt{\text{AM} * \text{HM}}$$

If a and b are two positive numbers:

$$\text{AM} = \frac{a+b}{2} \quad \text{GM} = \sqrt{ab} \quad \text{HM} = \frac{2ab}{a+b}$$

$$\text{HM} = \frac{\text{GM}^2}{\text{AM}} \quad \text{or GM} = \sqrt{\text{AM} * \text{HM}}$$

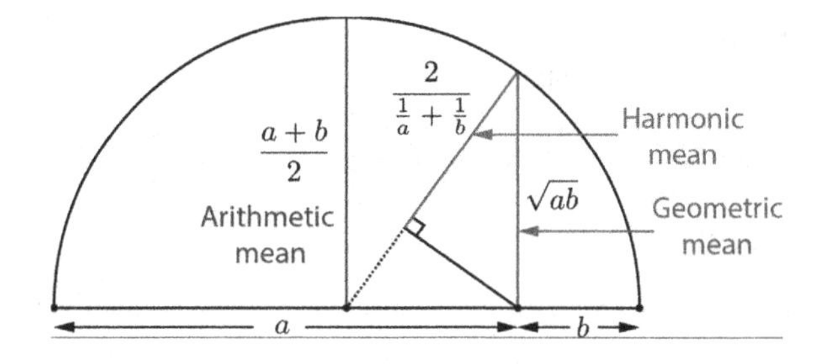

Fig. 2.6 Relationship between AM, HM and GM

Example 21 Given that $a > b$, find the harmonic mean of these two numbers if their arithmetic mean is 16 and geometric mean is 8.

Solution
$\text{HM} = \frac{\text{GM}^2}{\text{AM}} = \frac{64}{16} = 4.$

2.2 Partition Values

As the name suggests, partition values or fractals refer to dissecting the data into equal parts. When we split the data set and inspect each of them to derive some meaning/ or to examine the data, we get fractals. Types of partition values are based on the number of equal parts they are divided into. A few popular ones are, quartiles, deciles and percentiles.

i. **Quartiles**—Data is equally divided **in**to 4 equal parts with 3 divisions, and each unit is named as, Q_1, Q_2 and Q_3.
i. **Deciles**—Data is equally divided **in**to 10 equal parts with 9 divisions, and each unit is named as, D_1, D_2, ..., D_3.
ii. **Percentiles**- Data is equally divided **in**to 100 equal parts with 99 divisions, and each unit is named as, P_1, P_2, ..., P_3.

Example 22 $\mathbf{D_6}$: When data is divided into ten equal parts, this is the sixth part and is positioned after 60% of data.

$\mathbf{P_{80}}$: When data is divided into 100 equal parts, this is the 80th part of data.

$\mathbf{Q_2}$: This is the second quartile value which is also called the median value.

Note: These are positional values. They only denote where the values in the data set lie. We need to identify the value corresponding to their position in the data given to us.

There are 3 types of calculations (or formulae) for partition values based on the data type.

- Raw data.
- Discrete data and,
- Continuous data.

Calculations of Partition Values

Raw Data

For any discrete data set, follow these steps to calculate the partition values.

Step 1: Arrange data in an orderly manner, generally ascending values.

Step 2: Plug in the desired values into this formula. This step will give us the location at which the partition values exist.

ith position: $\frac{\text{i(n+1)}}{\text{Total divisions}}$, n-number of elements and **i:** positional value.
Total Divisions in case of,

- Quartiles = 4, deciles = 10 and percentiles = 100.

Step 3: The value or average values at the result of step 2 will be the desired solution.

For example:

P_{86} The total division is 100, $i = 86$.
D_6 The total division is 10 and $i = 6$.
Q_1 The total division is 4 and $i = 1$.

Example 23 Calculate all the quartile values for the data set:
5, 71, 54, 98, 43, 76, 1, 34, 66 and 21.

Solution
Data arranged in ascending order: 1, 5, 21, 34, 43, 54, 66, 71, 76 and 98.
In this problem, the values of **i** = 1, 2 and 3. Total divisions = 4 (Quartiles) and $n = 10$. Therefore,

$$Q_1 = \left(\frac{n+1}{4}\right)^{\text{th}} \text{term.}$$

$$Q_2 = 2\left(\frac{(n+1)}{4}\right)^{\text{th}} = \left(\frac{(n+1)}{2}\right)^{\text{th}} \text{term.}$$

$$\text{and} \quad Q_3 = 3\left(\frac{n+1}{4}\right)^{\text{th}} \text{term.}$$

Now locating the corresponding values.
$Q_1 = \frac{i(n+1)}{4} = \frac{11}{4} = 2.75$ which means, the value of the first quartile at 2.75th position is:

$$Q_1 = \text{2nd term} + 0.75 * (\text{3rd term} - \text{2nd term})$$
$$Q_1 = 5 + \{0.75 * (21 - 5)\}$$
$$Q_1 = 5 + 12$$
$$Q_1 = \mathbf{17}.$$

$Q_2 = \frac{i(n+1)}{4} = \frac{2(11)}{4} = 5.5$ which means, the value of the second quartile at 5.5th position is:

$$Q_2 = \text{5th term} + 0.5^*(\text{6th term} - \text{5th term})$$
$$Q_2 = 43 + \{0.5^*(54 - 43)\}$$
$$Q_2 = 43 + 5.5$$
$$Q_2 = \mathbf{48.5}$$

$Q_3 = \frac{i(n+1)}{4} = \frac{3(11)}{4} = 8.25$ which means, the value of the third quartile at 5.5th position is:

$$Q_3 = \text{8th term} + 0.25 * (\text{9th term} - \text{8th term})$$
$$Q_3 = 71 + \{0.25 * (76 - 71)\}$$
$$Q_3 = 71 + 1.25$$
$$Q_3 = \mathbf{72.25}$$

Example 24 Calculate all the decile values D_1, D_3, D_7 and D_9 for the data set, 5, 71, 54, 98, 31, 43, 76, 1, 34, 66 21,9,70, 99, 47, 55, 2, 95, 63 and 72.

Solution

Data arranged in ascending order:

1, 2, 5, 9, 21, 31, 34, 43, 47, 54, 55, 63, 66, 70, 71, 76, 95, 98 and 99.

In this problem, the values of $\boldsymbol{i} = 1, 3, 7$ and 9. Total divisions = 10 (Deciles) and n = 19. Therefore,

$$D_1 = \frac{i(n+1)}{10} = \frac{1(19+1)}{10} = \frac{20}{10} = 2$$

This means the value of the decile at the 2nd position is **2**.

$$D_3 = \frac{i(n+1)}{10} = \frac{3(19+1)}{10} = \frac{60}{10} = 6.$$

This means the value of the decile at the 6th position is **31**.

$$D_7 = \frac{i(n+1)}{10} = \frac{7(19+1)}{10} = \frac{140}{10} = 14.$$

This means the value of the decile at the 14th position is **70**.

$$D_{10} = \frac{i(n+1)}{10} = \frac{10(19+1)}{10} = \frac{180}{10} = 18.$$

This means the value of the decile at the 18th position is **99**.

Example 25 Calculate all the decile values P_{10}, P_{30} and P_{91}.

For the data set: 3, 1.8, 9, 5.6, 21, 29, 43, 34.15, 31.9, 44.7, 55.8, 63, 54, 66, 65.5, 71, 70.44, 98, 76.25, 80.64, 88.94, 73.25 and 94.63.

Table 2.18 Discrete data set for calculation of partition values

Sl. No.	Value	Sl. No.	Value
1	1.8	13	63
2	3	14	65.5
3	5.6	15	66
4	9	16	70.44
5	21	17	71
6	29	18	73.25
7	31.9	19	76.25
8	34.15	20	80.64
9	43	21	88.94
10	44.7	22	94.63
11	54	23	98
12	55.8		

Solution
In this problem, the values of $\boldsymbol{i} = 10$, 30 and 91. Total divisions = 100 (percentiles) and $n = 23$. Therefore, the data arranged in ascending order is (Table 2.18).

$$P_{10} = \frac{10(23+1)}{100} = \frac{240}{100} = 2.4$$

which means, the value of tenth percentile at 2.4th position is:

$$P_{10} = \text{2nd term} + 0.4 * (\text{3rd term} - \text{2nd term})$$
$$P_{10} = 3 + \{0.4^*(5.6 - 3)\}$$
$$P_{10} = 3 + 1.04$$
$$P_{10} = \mathbf{4.04}$$

$$P_{30} = \frac{30(23+1)}{100} = \frac{720}{100} = 7.2$$

This means the value of 30th percentile at 7.2th position is:

$$P_{30} = \text{7th term} + 0.2 * (\text{8th term} - \text{7th term})$$
$$P_{30} = 31.9 + \{0.2 * (34.5 - 31.9)\}$$
$$P_{30} = 31.9 + 2.6$$
$$P_{30} = \mathbf{34.5}$$

$$P_{91} = \frac{91(23+1)}{100} = \frac{2184}{100} = 21.84$$

This means the value of 91st percentile at 21.84th position is:

$$P_{91} = 21\text{st term} + 0.84 * (22\text{nd term} - 21\text{st term})$$
$$P_{91} = 88.94 + \{0.84 * (94.63 - 88.94)\}$$
$$P_{91} = 88.94 + 4.77$$
$$P_{91} = \mathbf{93.71}$$

Discrete frequency distribution:

Step 1: Obtain the cumulative frequencies.
Step 2: Calculate, $\frac{i(N+1)}{\text{Total Divisions}}$ value where N is the total frequency and total divisions will be 4 for quartiles, 10 for deciles and 100 for percentiles.
Step 3: The value of the variable corresponding to the first cumulative frequency greater than or equal to the result obtained in step 2 is the position of the partition value.

Example 26 In a village, the village panchayat has planned to conduct a health check-up for the heads of the family. This table shows the data concerning the specific age that the health camp has requested. Calculate the values of $\mathbf{Q_3}$, $\mathbf{D_9}$ and $\mathbf{P_{95}}$ for this particular data.

Age (years)	No. of family members
25	61
35	135
45	158
55	190
65	50
75	35
85	10
Total	639

See (Table 2.19).

Solution

$$Q_3 = \frac{3 * (639 + 1)}{4} = \frac{1920}{4} = 480$$

The corresponding value of age that is greater than 480 in the cumulative frequency column is **55**. Therefore,

$$Q_3 = 55$$

Table 2.19 Cumulative frequency distribution of number of families age wise

Age (years)	No. of family members	Cumulative frequency
25	61	61
35	135	196
45	158	354
55	190	544
65	50	594
75	35	629
85	10	639
Total	639	3017

$$D_9 = \frac{9 * (639 + 1)}{10} = \frac{5760}{10} = 576$$

The corresponding value of age that is greater than 576 in cumulative frequency column is **65**. Therefore,

$$D_9 = 55$$

$$P_{95} = \frac{95(640)}{100} = \frac{60800}{100} = 608$$

The corresponding value of age that is greater than 608 in the cumulative frequency column is **75**. Therefore, $P_{95} = 608$.

Continuous frequency distribution:

Step 1: Check whether the class intervals are of exclusive type. If not, convert the inclusive type class intervals to an exclusive type of class interval.
Step 2: Calculate the cumulative frequencies and also calculate $i(N/d)$ value.
Step 3: Identify the first cumulative frequency greater than or equal to $i(N/d)$ value. The corresponding class interval is the *i-quartile* class.
Step 4: Calculate the position using the below formula, where ***l*** is the lower limit of the ith quartile class, **N** is the total frequency, d is the division type (which is 4 in case of quartiles, 10 in case of deciles and 100 in case of percentiles), ***c*** is the cumulative frequency preceding the ith quartile class, ***f*** is the frequency of the *i*th quartile class and ***h*** is the width of the *i*th quartile class.

$$\text{Partition value} = 1 + \left\{ \frac{\left\{\left(i\frac{N}{d}\right) - c\right\}}{f} \right\} * h$$

Note: The class interval corresponding to the cumulative frequency which is immediately greater than the $i(N/d)$ value is the interval that holds the value of the required partition value.

Example 27 In an apple orchard, a survey was conducted in weights of apples grown after a chemical alteration was done to the soil. The tabulated data is provided. Calculate the, Q_3, D_4 and P_{10} and P_{99}values. To help the researcher figure out if there was any difference in the production of apples.

Weight of apples (grams)	Number of apples (f)
55–65	15
65–75	19
75–85	21
85–95	28
95–105	20
105–115	10
115–125	5

Solution

See Table 2.20.

The formula to calculate the partition values is,

$$\text{Partition value} = 1 + \left\{ \frac{\left\{\left(i\frac{N}{d}\right) - c\right\}}{f} \right\} * h$$

The corresponding class interval where the cumulative frequency is just higher than $i * (N/d)$ value is expected to have the partition value. Where l is the lower limit, h is height, f is the frequency of the class interval that contains the partition value. Whereas, c is the cumulative frequency of the preceding class interval.

Table 2.20 Cumulative frequency table for weights of apples

Weight of apples (grams)	Number of apples (f)	Cumulative frequency
55–65	15	15
65–75	19	34
75–85	21	55
85–95	28	83
95–105	20	103
105–115	10	113
115–125	5	118
Total	118	521

$$Q_3 = \frac{iN}{4} = \frac{3 * 118}{4} = 88.5$$

The value at 88.5 lies in the class interval 95–105. Therefore, the $l = 95, f = 20$, $h = 10$, $c = 83$.

$$Q_3 = 95 + \left\{\frac{88.5 - 83}{20}\right\} * 10$$

$$Q_3 = 95 + 2.65 = \mathbf{97.65}$$

$$D_4 = \frac{iN}{10} = \frac{4 * 118}{10} = \frac{472}{10} = 47.2$$

The value at 47.2 lies in the class interval 75–85. Therefore, the $l = 75, f = 21$, $h = 10$, $c = 34$.

$$D_4 = 75 + \left\{\frac{47.2 - 34}{21}\right\} * 10$$

$$D_4 = 75 + \left\{\frac{13.2}{21}\right\} * 10$$

$$D_4 = 75 + 6.28 = \mathbf{81.28}$$

$$P_{10} = \frac{iN}{100} = \frac{10 * 118}{100} = 11.8$$

The value at 11.8 lies in the class interval 55–65. Therefore, the $l = 55, f = 15$, $h = 10$, $c = 0$ (This is the class interval with no previous cumulative frequency.)

$$P_{10} = 55 + \left\{\frac{11.8 - 0}{15}\right\} * 10$$

$$P_{10} = 55 + 7.86 = \mathbf{62.86}$$

$$P_{99} = \frac{iN}{100} = \frac{99 * 118}{100} = 116.82$$

The value at 116.82 lies in the class interval 115–125. Therefore, the $l = 115, f = 5$, $h = 10$, $c = 113$,

$$P_{10} = 115 + \left\{\frac{116.82 - 113}{5}\right\} * 10$$

$$P_{10} = 115 + 7.64 = \mathbf{122.64}$$

Case Study 1

In certain studies, unlike physical sciences, measurements are not absolute and exact. This concept was dealt with in-depth while discussing scales. Therefore, raw data does not reveal much and there is a need to use some other appropriate technique to conclude. A score of 70 in an achievement test has no meaning unless it is interpreted in terms of the reference group. When the number of students who have scored equal to, less than and more than this score is found out, meaningful interpretations can be drawn. This involves the use of percentiles.

Example 28 Two girls studying in two different colleges get scores 40 and 55 on a statistics test. If we just look at the numbers, the conclusion could be that the second student got higher marks, which is true. But if we calculate the percentile of their scores in their class and suppose both are at 80 percentiles, we can conclude that, in their respective classes, they are equal in terms of their achievements.

Case Study 2

A Study titled "**How to analyse percentile impact data meaningfully in bibliometrics: The statistical analysis of distributions, percentile rank classes and top-cited papers**" by Lutz Bornmann is reproduced here:

"According to current research in bibliometrics, percentiles (or percentile rank classes) are the most suitable method for normalizing the citation counts of individual publications in terms of the subject area, the document type and the publication year. Up to now, bibliometric research has concerned itself primarily with the calculation of percentiles. This study suggests how percentiles can be analysed meaningfully for an evaluation study. Publication sets from four universities are compared with each other to provide sample data. These suggestions take into account, on the one hand, the distribution of percentiles over the publications in the sets (here: universities) and on the other hand concentrate on the range of publications with the highest citation impact—that is, the range which is usually of most interest in the evaluation of scientific performance."

Right tool: Percentiles

Reason: For a researcher, the number of times his/ her paper is cited is more important than the total number of papers produced. This holds good for universities and research institutions as well. Percentiles help in measuring this and if, for instance, the researcher is in the high citation of 1%, it means his work is discussed and used more in research fields.

Case Study 3

Consider a test score that is not for just evaluation but also for comparison. Many students will cluster into the median; i.e., their percentiles will be between 25 and 75, whereas some may reach the 80–90 s range. The average and median scores can be computed into expected results and will show the performance of people as well as how each individual is performing.

Case Study 4

Consider a test in which every student taking a test answers the same question incorrectly or if the majority answer it wrongly, (this can be found using percentiles) then it needs to be checked. Either the data or wordings are wrong or maybe the concept was not covered properly in the class. This can help to standardize and maintain quality in testing students.

Case Study 5

Consider a rural school without proper facilities for teaching. If the whole school scores well below average in academic tests, that is the students fall in the low percentile category, this situation needs attention. The students may have studied in the local language and may not be able to attempt tests in other languages, or the tutor has not done an efficient job or some other issues may be there.

Case Study 6

Consider a study of air samples taken from an urban area for their sulphur dioxide content in parts per million (ppm). If the safe level is defined as 3 ppm and the percentile for 3 ppm is 35%, then it can be said that 65% of samples exceed the safe limit and that action is required to reduce pollution levels.

Case Study 7

[7] Consider a college profile that presents the following SAT scores for the 25th and 75th percentiles:

- SAT Math: 540/620
- SAT Writing: 490/600
- SAT Critical Reading: 500/610

(The lower number is for the 25th percentile of students who enrolled in the college and the upper number is for the 75th percentile of students who enrolled in the college.)

Meaning, 25% of enrolled students received a math score of 540 or lower. 75% of enrolled students got a math score of 620 or lower (in other words, 25% of students got above 620). For the school above, if one has an SAT math score of 640, that person would be in the top 25% of applicants. Similarly, if one has a math score of 500, that person is in the bottom 25% of applicants.

Right tool: Percentiles

Reason: SAT scores are used for evaluation and admission purposes. It is more important to see the relative ranking of a student rather than his marks in isolation as he/ she is competing with other students (Table 2.21).

Example 29 Here is the data for admissions for Ivy League Colleges in the USA.

Right tool: Percentiles

Table 2.21 SAT scores of students in Math, Writing and Reading

SAT scores						
	Reading		Math		Writing	
	25%	75%	25%	75%	25%	75%
Brown	630	740	650	760	640	750
Columbia	690	780	700	790	690	780
Cornell	630	730	670	770	–	–
Dartmouth	670	780	690	790	690	790
Harvard	690	790	700	800	690	790
Princeton	700	790	710	800	700	790
U Penn	660	750	690	780	670	770
Yale	700	800	710	790	710	800

Reason: The above case studies indicate how important percentiles are. Here, the individual scores do not have a meaning by themselves but are used to calculate percentiles and then to draw conclusions.

2.3 Measures of Dispersion

By dispersion, we mean the extent to which data is scattered. In Fig. 2.7, we can notice the levels of dispersion. Therefore, to define this spread of data we use statistical tools called, measures of dispersion. These are broadly categorized into 2 groups. Absolute measures and relative measures of dispersion.

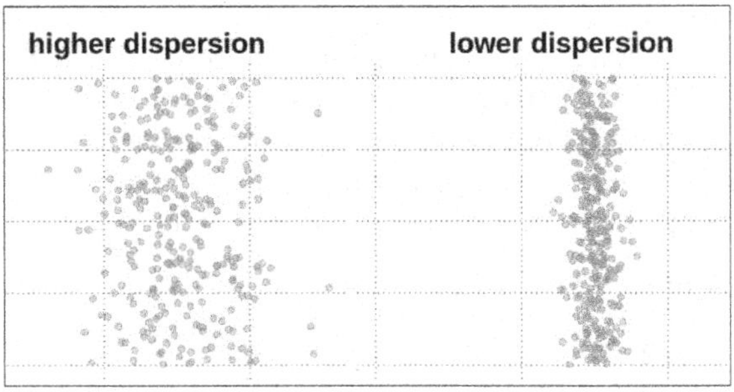

Fig. 2.7 Example of dispersion of data

Absolute Measures:

Absolute measures are those which measure the dispersion in the same units of the data, like for example in kilograms, miles and rupees. They are individual measures that are not used for comparisons. Absolute measures of dispersion are dependent on units of measurement and thus cannot be used for comparison if the groups have different units.

Example 30 Range, quartile deviation, mean deviation, standard deviation.

Relative Measures:

Relative measures are those which measure the dispersion of data which are independent of units of measurement. They are pure independent numbers, thus used for comparisons.

Example 31 Coefficient of range, coefficient of quartile deviation, etc. (Fig. 2.8).

Range and Coefficient of Range

Definition

Range is the difference between the largest and the smallest observations in the data. If A and B are the largest and smallest observations in the data, respectively, then the range is given by:

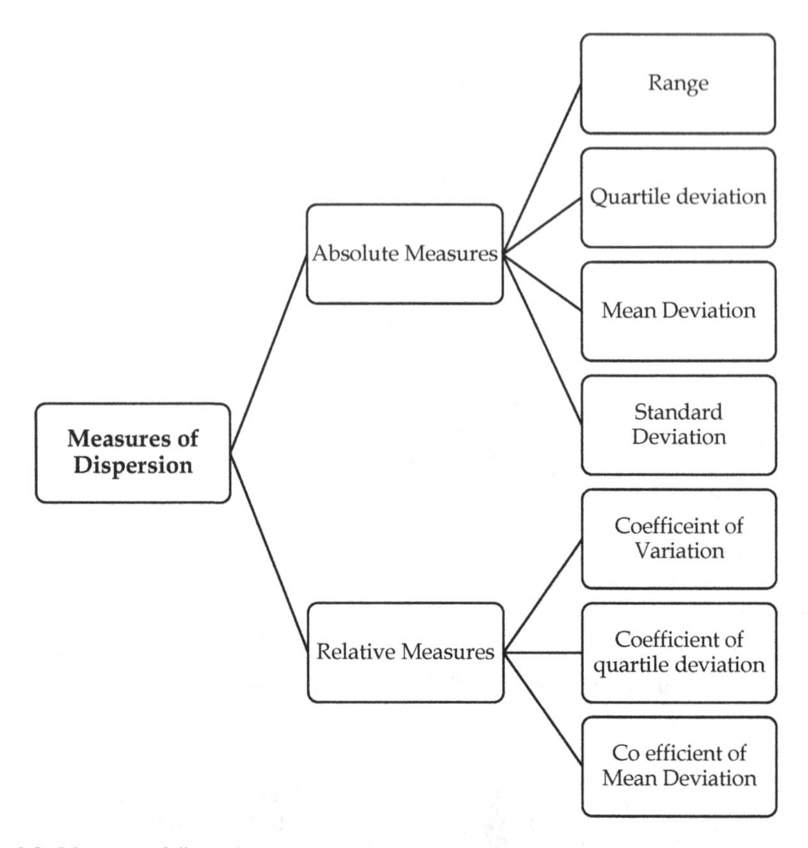

Fig. 2.8 Measures of dispersion

$$\mathbf{Range} = \mathbf{X_{Max}} - \mathbf{X_{Min}}$$

The range is the simplest but crude measure of dispersion. It is not at all a reliable measure of dispersion as it is based on only two extreme values which themselves are subject to chance fluctuations.

Coefficient of Range is given as

$$\frac{X_{\max} - X_{\max}}{X_{\max} + X_{\max}} = \frac{A - B}{A + B}.$$

Example 32 Suppose a researcher is measuring a variable that has to lie within a certain limit. Let this variable be the age of school children. If he wants to check quickly about the correctness of data entry, he/she has to calculate the range. If it

Table 2.22 Weights of articles produced by Machine A and Machine B

Machine A	125	100	105	109	111	104	113	120	108
Machine B	112	108	111	127	102	104	117	100	119

comes out to be, say 145 years, then obviously, either the upper limit or lower limit has been entered incorrectly.

Example 33 A researcher studying volatility in the stock market will study the range. Large ranges indicate high volatility and small ranges indicate low volatility. No other measure can so easily give such an indicator.

Case Study 1

Two machines produce articles whose weights (grams) are given below. The researcher wants to determine if the machines are producing articles conforming to prescribed standards (Table 2.22).

Right tool: Range

Reason: A quick review can be done using the range. It takes time to calculate other measures of dispersion. A comparison of machines A and B also becomes possible.

Calculation:

The range of weights of articles produced by Machine A = 125–104 = 21.

Coefficient of range, Machine A = **0.09**.

The range of weights of articles produced by Machine B = 127–100 = 27.

Coefficient of range, Machine B = **0.11**.

Thus, Machine B has a higher range and it can produce articles over a wider range.

Case Study: The following data gives the weight (pounds) of 18 men who were on the crew teams at Oxford and Cambridge universities (Hand et al. 1994, p. 337). Suppose the researcher wishes to find out whether a person with a particular weight can get on the team? (Table 2.23)

Right tool: Range

Reason: A look at the range will reveal immediately if the person's height is in the range of weights of people in the team.

Table 2.23 Weights of men from Oxford and Cambridge

Cambridge	188.5	183.0	194.5	185.0	214.0	203.5	186.0	178.5	109.0
Oxford	186.0	184.5	204.0	184.5	195.5	202.5	174.0	183.0	109.5

Table 2.24 List of women having their first child at various ages

Age	Under 20	21–24	25–29	30–34	35 and above
Percentage of women	24	34	25	15	2

Calculation: Range of weights of men who were on the crew team at Cambridge = 214.0–109.0 = 105.

Coefficient of range, Cambridge = **0.325**.

Range of weights of men who were on the crew team at Oxford = 204.0–109.5 = **94.5**.

Coefficient of range, Oxford = **0.301**.

The Cambridge team has a wider range so a person desirous of getting into the Cambridge team has a better chance.

Quartile Deviation QD (Interquartile Range)

Definition
The interquartile deviation QD is given by:
$\text{QD} = \frac{1}{2}(Q_3 - Q_1)$ where Q_1 and Q_2 are the first and third quartiles of the distribution respectively. This helps to overcome the problem of range, namely dependence on extreme values. It plays an important role in Box plots.

Case study 1
Consider the following data of the percentage of women having their first child at various ages. Which measure of dispersion should the researcher use? [8] (Table 2.24).

Right tool: Quartile deviation

Reason: Range will give only a partial picture and since the class intervals are open-ended; we can only assume the lower limit of the first class and the upper limit of the last class. This makes the range very vague. Also, due to the open-ended class intervals, standard deviation (which will be explained ahead in this chapter) cannot be computed.

Solution
To convert an exclusive class interval into an inclusive class interval:

> **Step 1**: Find the difference between the lower limit of a class and the upper limit of the previous class interval.
> **Step 2**: Divide this difference between 2. Meaning, take the arithmetic average of the limits.
> **Step 3**: This difference is added to the upper limit of the first-class interval and then reduced from the lower limit of the second-class interval.

Table 2.25 Converting ages of women from exclusive to inclusive class interval

Age	Percentage of women	Age (exclusive class interval)	Mid-point	Cumulative frequency
Under 20	24	Under 20.5	–	24
21–24	34	20.5–24.5	22.5	58
25–29	25	24.5–29.5	27	83
30–34	15	29.5–34.5	32	98
35 and above	2	34.5 and above	–	100

Step 4: Repeat this process with every consecutive class interval. This makes the class intervals inclusive and also continuous.

Example 34 Like in the above example, we have converted the class intervals into inclusive so that we can compute the quartiles. The difference between the upper limit of first-class interval-20 and the lower limit of second-class interval 21 is 1. So, the average is 0.5, which is added to 20 and reduced from 21. Like this, we made the distribution table with inclusive class intervals, like how it's shown in Table 2.25

Quartiles and Coefficient of Quartiles

$$Q_1 = l + \left\{ \frac{i\left(\frac{N}{d}\right) - cf}{f} \right\} * h$$

$$Q_1 = 20.5 + \left\{ \frac{1\left(\frac{100}{4}\right) - 24}{34} \right\} * 5$$

$$Q_1 = 20.5 + \left\{ \frac{1}{34} \right\} * 5$$

$$Q_1 = 20.5 + 0.147 = \mathbf{20.647}$$

$$Q_3 = l + \left\{ \frac{3\left(\frac{100}{4}\right) - cf}{f} \right\} * h$$

$$Q_3 = 24.5 + \left\{ \frac{75 - 58}{25} \right\} * 5$$

$$Q_3 = 24.5 + \left\{ \frac{17}{25} \right\} * 5$$

$$Q_3 = 24.5 + 3.4 = \mathbf{27.9}$$

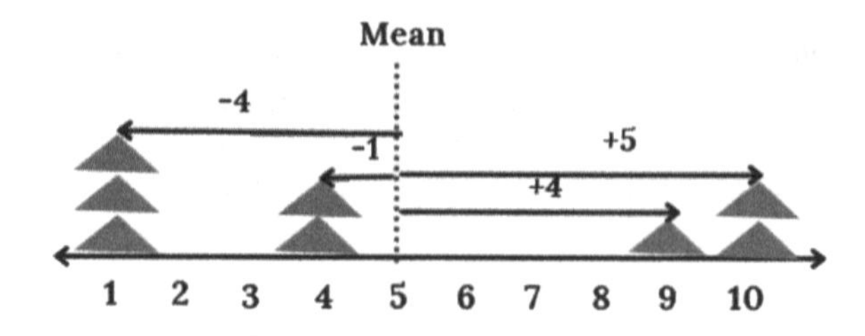

Fig. 2.9 Mean deviation

$$\text{Coefficient of Quartiles} = \frac{26.9 - 20.647}{2} = \frac{6.253}{2} = \mathbf{3.12}$$

Mean Deviations Around a Central Tendency

(Mean deviation from an average, median and mode.)

Deviation in statistics measures, "how much the value deviates from the central measure"? The most commonly used central tendency to calculate the deviation is around the mean. A mean deviation is a statistical tool that measures the average deviation of data from the mean which is here considered as a central value. Deviations from the median and mode can also be calculated based on the use case.

Consider Fig. 2.9. The data set represented here is 1, 1, 1, 4, 4, 9, 10 and 10. The arithmetic mean is 5. (Sum of all observations/total number of observations). Deviation refers to the distance between the data points and the mean which is constant and, in this case, it is 5. So, the differences are as follows.

Between 1 and 5 $= 1 - 5 = -4$. (2 times)
Between 4 and 5 $= 4 - 5 = -1$.
Between 9 and 5 $= 9 - 5 = +4$ and
Between 10 and 5 $= 10 - 5 = +5$ (Thrice).

Mean deviation is nothing but the average of these 8 distance points that are deviated from the mean. $(-4) + (-4) + (-4) + (-1) + (-1) + 4, +5, +5$. Whereas the sum of all the deviations from the mean is generally 0. That is because of the property of the arithmetic mean. It is positioned in such a way that the value of deviations above the mean nullifies the value of deviations below the mean. Let's see what the deviations around mode and median will look like and how they can help us in data analytics.

Definition

Raw data: Let x_i, $i = 1, 2, \ldots, n$ be $\boldsymbol{n}$ observations, then the deviation from the average A (usually mean, median or mode) is given by:

Short form of deviation from average written as **D**:

$$D = \frac{1}{n}\sum_{i=1}^{n}|x_i - A|$$

where A is the central value from which the distance of observations is to be calculated.

Grouped data:

If $x_i | f_i, i = 1, 2, \ldots, n$ is the frequency distribution, then the deviation from the average A (usually mean, median or mode) is given by:

$$D = \frac{1}{N} \sum_{i=1}^{n} f_i |x_i - A|, \sum_i f_i = N$$

where $|x_i - A|$ represents the absolute value of the deviation $(x_i - A)$

Case Study 1

Hand et al. (1994, p. 148) provide data on the number of words in each of 600 randomly selected sentences from the book "Shorter history of England" by G. K. Chesterton. They summarized the data as follows. If the researcher wants to know how much is the median and mode and also the dispersion around it, which is the most suitable measure?

Number of words	Frequency	Number of words	Frequency
1–5	3	31–35	68
6–10	27	36–40	41
11–15	71	41–45	28
16–20	113	46–50	18
21–25	107	51–55	12
26–30	109	56–60	3

Calculation:

See Table 2.26.

The maximum frequency in the second column is 113, and thus the modal class is 16–20. Therefore, the highlighted row contains the precise modal value.

The value to plug into the formula is: $1 = 16$, $f_1 = 113$, $f_2 = 107$ and $f_0 = 71$, $h = 5$.

$$\text{Mode} = 1 + \left\{ \frac{f_1 - f_0}{2f_1 - f_0 - f_2} \right\} * h$$

$$\text{Mode} = 16 + \left\{ \frac{113 - 71}{(2 * 113) - 71 - 107} \right\} * 5$$

$$\textbf{Mode} = \textbf{20.37}$$

Now, let's calculate the deviations from mode (Table 2.27). Which is,

Table 2.26 Table calculating the dispersion around mode for words collected at random

Number of words	Frequency		Mid-point (X)		
1–5	3		2.5	17.87	53.61
6–10	27		7.5	18.73	505.71
11–15	71		12.5	13.73	974.83
16–20	**113**		**17.5**	**8.73**	**986.49**
21–25	107		22.5	3.73	399.11
26–30	109		27.5	1.27	138.43
31–35	68		32.5	6.27	426.36
36–40	41		37.5	11.27	462.07
41–45	28		42.5	16.27	455.56
46–50	18		47.5	21.27	382.86
51–55	12		52.5	26.27	315.24
56–60	3		57.5	31.27	93.81
Total	600		360	176.68	5194.08

$$\frac{\sum |\mathbf{f}^*(\mathbf{x_i} - \mathbf{Mode})|}{\mathbf{N}} = \frac{5194.08}{600} = \mathbf{8.658}$$

To calculate the median of the data, we divide the data set into half and we get 600/2 = 300. In the cumulative frequency column, the immediate value that is just greater than 300 is 321, and therefore the corresponding class interval is 21–25. (The highlighted row.) Since the class interval 21–25 is expected to have the median value

Table 2.27 Table calculating the dispersion around median for words collected at random

Number of words	Frequency	Mid-point (X)	Cumulative frequency	$\lvert X_i - \text{Median}\rvert$	$\lvert f^*(x_i - \text{Median})\rvert$
1–5	3	2.5	3	22.5	67.5
6–10	27	7.5	30	17.5	472.5
11–15	71	12.5	101	12.5	887.5
16–20	113	17.5	214	7.5	847.5
21–25	**107**	**22.5**	**321**	**2.5**	**267.5**
26–30	109	27.5	430	2.5	272.5
31–35	68	32.5	498	7.5	510
36–40	41	37.5	539	12.5	512.5
41–45	28	42.5	567	17.5	490
46–50	18	47.5	585	22.5	405
51–55	12	52.5	597	27.5	330
56–60	3	57.5	600	32.5	97.5
Total	600	360	4485	185	5160

in it, we locate the value using the formula,

$$\text{Median} = l + \left\{ \frac{\frac{N}{2} - c}{f} \right\} * h$$

where $l = 21$, $c = 214$ (Cumulative frequency of the previous class interval), $f = 107$ and $h = 5$. Therefore,

$$\text{Median} = 21 + \left\{ \frac{300 - 214}{107} \right\} * 5 = 21 + 4.01 = \mathbf{25.01}$$

This value of median is greater than the maximum of the class interval 25, but also note that these class intervals are of an inclusive type and when we convert to exclusive class interval type the median class is in the range from 20.5 to 25.5 and the successive class interval will be 25.5–29.5. This means the median value slightly higher than 25 is still included in the class interval that is highlighted in the row.

To calculate the deviations around the median.

$$\text{Mean deviation around median} = \frac{\sum |f * (x_i - \text{Median})|}{N} = \frac{5160}{600} = \mathbf{8.6}$$

In this example, we see that the deviation around the median which is 8.6 and around the mode is 8.65. This difference is very minimal.

Case Study 2

Langlois and Roggman (1990) conducted a study on the perceived attractiveness of faces. They presented students with computer-generated pictures of faces; some of them were created by averaging snapshots of different people. They had created different sets of photographs by averaging 4 snapshots in one set and 32 in another. Students were asked to rate each one on a 5-point scale of attractiveness. The researchers were primarily interested in determining if the differences in the mean rating of faces in the first set were less than the mean rating in the second set.

Right tool: Average deviation around the mean value.

Reason: When only 4 faces are averaged, there is room for individuality whereas when 32 faces are averaged, the result becomes very different. In this case, the researchers are interested in how much the scores deviate from the mean without bothering about the direction, that is less than or more than the mean.

2.4 Understanding Standard Deviation and Variance

Definition
Standard deviation is the positive square root of the average of squared deviations of observations from the arithmetic mean. It is denoted by the symbol σ. Variance is the square of standard deviation. Both these concepts are extensively used concepts in statistical analysis (Fig. 2.10).

To understand better: Assume you are a dienophile, (a person who has a keen interest in dinosaurs) and you wish to measure the heights of different dinosaurs.

The heights of the 4 dinosaurs in Fig. 293.1 are 450, 780, 200 and 650 and inches, respectively. Let's understand the concept of variance first, Variance is the average square of deviations from the mean. Therefore,

$$\text{The average height of the dinosaurs} = \frac{450 + 780 + 200 + 650}{4} = \frac{2080}{4} = 520\,\text{inches}$$

Variance is calculated as the average of squares of all the individual deviations around the mean. Which is,

$$\frac{(450 - 520)^2 + (780 - 520)^2 + (200 - 520)^2 + (650 - 520)^2}{4}$$
$$= \frac{4900 + 67600 + 102400 + 16900}{4} = \frac{191800}{4} = \mathbf{47950}.$$

Standard deviation of height of dinosaurs $= \sqrt{47950} = \mathbf{218.97\ inches}$.

This is indeed a huge number. Thus, we compute standard deviation, which is simply the square root of the variance for easy understanding. Also, units of measurement are associated with standard deviation.

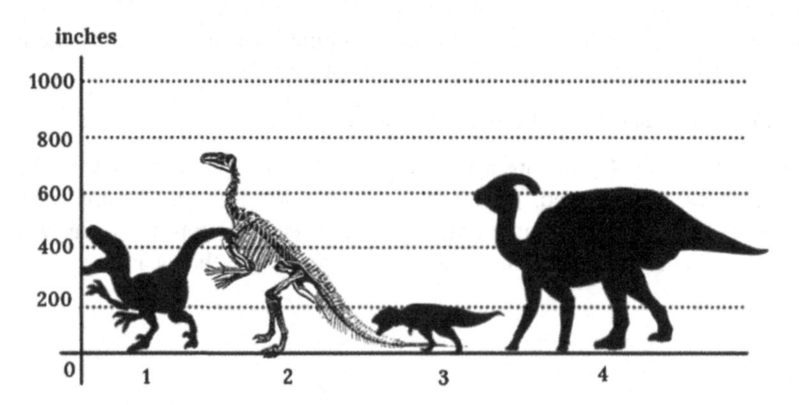

Fig. 2.10 Heights of types of dinosaurs

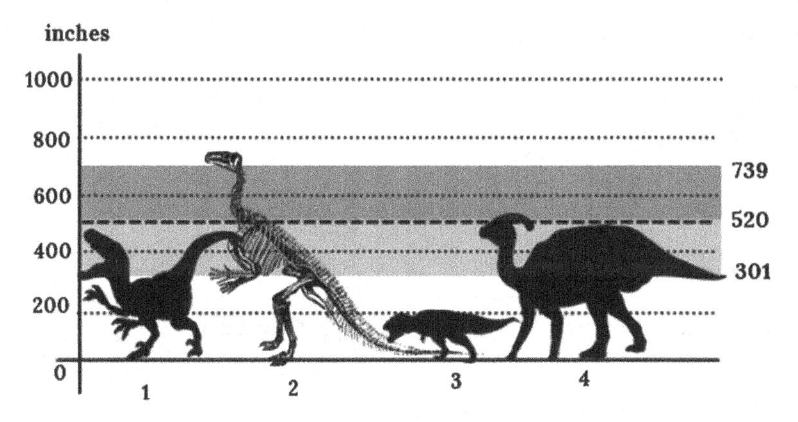

Fig. 2.11 Standard deviation of heights of dinosaurs

In this example, standard deviation as the name suggests, answers the question of "What is the normal height of dinosaurs? Which dinosaur is extra small and which is extra tall"?

What statisticians generally do with standard deviation is to add and reduce this value from the mean. Here in this example, we are showing the heights of dinosaurs under the 1 standard deviation limit.

The upper limit of the band = 520 + 219 inches (Rounding up) = 739 inches.

The lower limit of the band = 520 − 219 inches = 301 inches.

Take a look at Fig. 2.11. The shaded region in the chart shows the standard heights of dinosaurs. We create symmetry by adding and reducing a certain value from the mean. This value is the standard value of the deviations from the mean, hence called "**Standard Deviation**".

Definition

Standard deviation is the measure of the dispersion of a set of data from its mean. It measures the absolute variability of a distribution. The higher the dispersion or variability, the greater is the standard deviation and the greater will be the magnitude of the deviation of the value from their mean.

Note: Since this is a small sample of 4 units, we use just n in the denominator. But if the data is a **Sample** (a selection taken from a bigger population), we would use $n - 1$ for the denominator.

Inference:

- From these limits, we understand that Dinosaur 3 is the shortest when compared.
- Dinosaur 2 is the tallest of all but slightly taller than the others in the range, its height is not way too far or abnormal like the third Dino.

Standard deviation formula for:

Raw data:

Standard deviation, $\sigma = \sqrt{\frac{1}{n}\sum_{i=1}^{n}(x_i - \overline{x})^2}$.

Variance, $\sigma^2 = \frac{1}{n}\sum_{i=1}^{n}(x_i - \overline{x})^2$ where, $\overline{x} = \frac{1}{n}\sum_{i=1}^{n} x_i$.

Grouped data:

Standard deviation $= \sigma = \sqrt{\frac{1}{N}\sum_{i=1}^{k} f_i(x_i - \overline{x})^2}$**Variance** $= \frac{1}{N}\sum_{i=1}^{k} f_i(x_i - \overline{x})^2$

where, $\overline{x} = \frac{1}{n}\sum_{i=1}^{n} f_i x_i$ and $\sum_{i=1}^{k} f_i = N$.

The larger the standard deviation, the greater will be the spread of the data about its mean.

Understanding Variance

Consider an investment returning yields in three consecutive years which are 10, 15 and −11%. The average mean of the returns is 4.66%. The differences between each yield and the mean are 5.34, 10.34 and −15.66% for each successive year.

Generally, squaring each deviation will produce 28.515, 106.91 and 245.23%. After adding up the squared deviations, they give a total of 380.66%. By dividing the total additional sum by the number of values in the number set, the variance will be 126.88%. Calculating the square root of the variance will generate the standard deviation of **11.26%** of the investment returns. This value depicts the variations in the returns from the investment. The higher the variations, the more volatile the investments are. Lesser variations also indicate the stability of returns.

Variance, like standard deviation, measures the extent to which a set of numbers is spread out from the average or mean. But statistical analysts use variance to determine the deflection of a random variable from its standard value.

Large variability is an indication of a huge spread of values in the number set. However, a minimum variance illustrates the proximity of figures between each other and from the mean value. Identical values within a data set portray a zero variance. Similarly, every positive number indicates a nonzero variance since a square value cannot be negative.

Variance Versus Standard Deviation

The standard deviation shows the position of each value from the mean. Consider a value with unit measurement in metres. However, a variance is indicated in larger units such as metres squared while the standard deviation is expressed in original units such as metres.

Due to the larger unit values of expression, a variance number becomes harder to interpret, However, while making statistical analysis, a variance is preferred since it's more descriptive about the variability of the data than the standard deviation.

Properties of a Variance

- Variance cannot be negative because it's a squared value. It is either positive or zeroes. For example,
  $\mathbf{Var(X) \geq 0}$.
- The variance of a constant value is equivalent to zero. For an obvious reason that it's a constant having a fixed position and value.
  $\mathbf{Var}(k) = \mathbf{0}$
- Suppose k is a constant, when this is added to the variance, then it remains invariant. This is again because a constant value does not affect the variance.
  $\mathbf{Var(X + k) = Var(X)}$
- If the values are multiplied by a constant, the outcome of the variance is scaled by the square root of that constant. This is because multiplying every value in the data set will scale up the entire data set.
  $\mathbf{Var(kX) = k^2 Var(X)}$

When Data points are altered:
When a constant is added to every observation:

- The new sample mean is equal to the original mean plus the constant.
- The standard deviation is unaffected.

When every observation is multiplied by the same constant

- The new sample mean is equal to the original mean multiplied by the constant.
- The new sample standard deviation is equal to the original standard deviation multiplied by the magnitude of the constant.

2.5 Coefficient of Variation CV

During data analysis, we would like to compare a lot of data sets for better decision-making. The key statistical tool that analyses data on, consistency, uniformity, reliability and stability of the data- coefficient of variation or CV. This is the ratio of the mean and standard deviation of data expressed in terms of percentage.

This is a vital tool for comparing the degree of variation between two data sets. CV is inversely proportional to these concepts. This means, higher CV lesser consistency, reliability, stability and uniformity.

$$\text{Coefficient of variation} = \frac{\text{Standard deviation}}{\text{Mean}} * 100 = \frac{\sigma}{\overline{x}} * 100$$

CV is used mostly on a ratio scale as there is a true zero point. Thus, when a researcher has physical scales like height, weight, time, marks and others, this tool can be used. But when data is in interval scale, like mental ability, educational grades and achievement tests, the use of CV becomes debatable.

Case Study: A group of students in a class have a mean height of 165 cm with an SD of 14 cm. Their average weight is 68 kg with an SD of 7 kg. In which aspect are the students more variable?

Right tool: CV

Reason: Height and weight are ratio scale variables as they have an absolute zero. Thus, CV is the right measure for comparison.

Calculation: CV (height) = (14/165) * 100 = **8.4848%**

$$\text{CV(weight)} = (7/68) * 100 = \mathbf{10.2941\%}$$

Thus, the students are more vulnerable to variations in their body weight than their height.

Note: CV is not a suitable tool for comparing dependent data sets.

A test on mental fitness was given to students of a class, and the mean was 45 with SD 6. Suppose 10 new items which are relatively easy are added to the test. Suppose all of them do well. The mean now changes to 45 + 10 = 55. But as we know earlier, the SD remains constant.

Calculation:

$$\text{CV(before)} = (6/45) * 100 = \mathbf{13.3\%}$$
$$\text{CV(after)} = (6/55) * 100 = \mathbf{10.90\%}$$

Inference: This implies that CV can be altered with a change in mean and is **not a stable tool for use when data is in ratio scale**.

Contextual Cases

Case Study 1

The performance of a cricketer in home ground and overseas is given below. A researcher wishes to make a comparative study of the performance (Table 2.28).

Right tool: SD

Reason: Comparison of only means will not yield the correct results. The performance also depends on consistency, which can be measured using SD and CV.

Table 2.28 Performance of cricketers in home ground and over seas

Overseas	38	41	2	71	121	62	48	55	69
Home ground	72	88	26	133	12	77	42	102	83

Calculation:

Mean of overseas performance = **56.33**.

Standard deviation of overseas performance = **31.98**.

CV (Overseas performance) = (31.98/56.33) *100 = **56.77**.

Mean of home ground performance = **70.55**.

The standard deviation of home ground performance = **38.11**.

CV (Home ground performance) = (38.11/70.55) *100 = **54.01**.

Inference: The player is more consistent in his performance overseas when compared to the home ground but yet when just standard deviations are measured, it is found that deviation is more when played on the home ground even though the mean is higher in this case. This is a valuable conclusion that will help assess and analyse the phenomenon.

Note: An important application of CV is in the construction of %CV charts which are used in the textile industry. The variation among tensile strength measurements from the thin thread is significantly smaller than measurements taken from the heavy thread. This is due to the inherent physical properties of the fibre.

Combined standard deviation:

Let n_1 and n_2 be the sizes. $\overline{x}_1$ and $\overline{x}_2$ the means. σ_1 and σ_2 the standard deviations of two groups, respectively, then the standard deviation of the combined group is

$$\sigma = \sqrt{\frac{1}{n_1 + n_2}\left[n_1\left({\sigma_1}^2 + {d_1}^2\right) + n_2\left({\sigma_2}^2 + {d_2}^2\right)\right]}$$

where $d_1 = \overline{x}_1 - \overline{x}$, $d_2 = \overline{x}_2 - \overline{x}$ and $\overline{x} = \frac{n_1\overline{x}_1 + n_2\overline{x}_2}{n_1 + n_2}$

This is a property of SD that combined SD can be found for different groups. This is not possible for other measures of dispersion like range and quartile deviation. This algebraic property of SD helps in a large number of cases, especially while dealing with different groups like population divided into various strata, people divided according to gender and so on.

Case Study 5

The following data gives some data about ages (years) at which women got married. It is tabulated along with their educational status. Which measure of dispersion can give the picture of variation over the entire population? (Table 2.29)

Right tool: Combined SD

Reason: Combined SD is the appropriate measure as the SD of ages can be calculated separately and they can be combined. This is not possible for other measures of dispersion like range and QD. This algebraic property of SD helps in a large number

Table 2.29 Table representing mean and variance of women with different education status

Educational status	N	Mean	Variance
No schooling	350	15.5	22.8
Primary	320	17.2	20.22
High school	260	19.45	15.54
Senior secondary	340	22.6	15.01
Secondary	275	23.8	14
Degree	180	24	14.9

of cases, especially while dealing with different groups like population divided into various strata, people divided according to gender and so on.

Calculation: The formula above can be extended to k groups, and it becomes,

$$\sigma = \sqrt{\frac{1}{(n_1 + n_2 + \ldots + n_k)}\left[n_1\left({\sigma_1}^2 + {d_1}^2\right) + n_2\left({\sigma_2}^2 + {d_2}^2\right) + \cdots + n_k\left({\sigma_k}^2 + {d_k}^2\right)\right]}$$

$d_i = \overline{x}_i - \overline{x},\ i = 1, 2, \ldots, k$ and $\overline{x} = \frac{n_1\overline{x}_1 + n_2\overline{x}_2 + \cdots + n_k\overline{x}_k}{n_1 + n_2 + \cdots + n_k}$.

Using the above formula, the combined SD of the data above is **5.3269**.

Quick reference:

See Table 2.30.

Some common misconceptions about measures of central tendency and dispersion

1. The only three measures of central tendency are the mean, the median and the mode.

Table 2.30 Table representing use cases of various dispersion measures

Situation	Measure of dispersion
Require only the highest and the lowest score	Range
Scores are distributed normally/almost normally	Average deviation/standard deviation
Scores are skewed, containing a few extreme scores	Quartile deviation
The number of scores is very less	Range
We need to know the dispersion in less time and easily	Range
Standard deviation is influenced to a large extent by extreme values	Average deviation
Scores are truncated or irregular and are open-ended	Quartile deviation
A reliable measure of dispersion is needed	Standard deviation
Further analysis like correlation and the testing significance of differences in the mean is desired	Standard deviation

2. The word "average" is synonymous with the word "mean".
3. The mean is superior to either the median or the mode.
4. It is enough to mention the mean and standard deviation in a group of data.
5. Data which is approximately normal is superior to data that is skewed.
6. A standard deviation indicates how far a typical score will deviate from the mean.
7. If the male test scores have an SD = 10 and the female test scores have an SD = 20, the SD for the combined group of males and females will be equal to 15.
8. Mean and median measure the "typical" value in a set of data. This is not always true. It holds good only when the data values are clustered around mid-value.
9. Outliers can be determined by just observing the mean and saying how far it is from the mean.

Case Study: "Last year's severe odour problem was in part due to extreme weather conditions created in the woodland area by El Nino. The company official said woodland saw 170–189 pc of its normal rainfall. Excessive rain means water in the holding ponds takes longer to exit for irrigation, giving it more time to develop an odour [10] (Table 2.31).

Analysis: Let us consider the data for annual rainfall for Davis, California, in inches

Mean: 18.69 inches

Median: 16.72 inches

Range: 31.28 inches

For the year 1997–98, under discussion, rainfall was 29.69 inches. This is 170–180% of normal, which is correct, as the official said. But this value is within the range and so, it is not an outlier and thus the official's statement is not true.

Graphical summaries provide a visual value for the data and help in deciding about further inquiry into the data. The importance of descriptive statistics is that

Table 2.31 Table displaying rainfall across years

Year	Rainfall	Year	Rainfall	Year	Rainfall	Year	Rainfall
1951	20.66	1963	11.2	1975	6.14	1987	16.3
1952	16.72	1964	18.56	1976	7.69	1988	11.38
1953	13.51	1965	11.41	1977	27.69	1989	15.79
1954	14.1	1966	27.64	1978	17.25	1990	13.84
1955	25.37	1967	11.49	1979	25.06	1991	17.46
1956	12.05	1968	24.67	1980	12.03	1992	29.84
1957	28.74	1969	17.04	1981	31.29	1993	11.86
1958	10.98	1970	16.34	1982	37.42	1994	31.22
1959	12.55	1971	8.6	1983	16.67	1995	24.5
1960	12.75	1972	27.69	1984	15.74	1996	19.52
1961	14.99	1973	20.87	1985	27.47	1997	29.69
1962	27.1	1974	16.88	1986	10.81		

they are tools for interpreting and analysing data. Measures of central tendency can be directly compared to different years, and these measures for several years may be compared to one another in order to learn how the situation has changed. Measures of dispersion can also be useful in understanding which data values have abnormally high numbers.

2.6 Lorenz Curve and Gini Coefficient [11]

Lorenz Curve is an extensively used visual tool in Econometrics that demonstrates the distribution of wealth in an economy. The architect of this famous Lorenz curve is Max Otto Lorenz. In 1905, this American economist first mentioned the Lorenz curve in his undergraduate essay. Gini's index is a tool that helps in measuring the concentration of wealth (Fig. 2.12).

On the Axis:

X-axis—Percentiles of the population.
Y-axis—Cumulative income of people from different sections of the population.

- **Line of Equality**: The slant line of 45 degrees with slope 1 is called the **Line of Perfect**. **Equality of incomes** at this point the Gini coefficient is defined as 0. Any point on this line means, that everyone in the population earned the same salary, and thus everyone is treated equally.
- This line is like a standard line of comparison from which the inequality levels are measured. These curved lines beneath the straight lines explain the inequality.

Reading and Interpreting the Lorenz Curve

See Fig. 2.13.

In Fig. 2.14, there are two Lorenz curves defined. Have a look at the associated X-axis and Y-axis values on the top and bottom of the curve to interpret the graph.

Fig. 2.12 Max Otto Lorenz

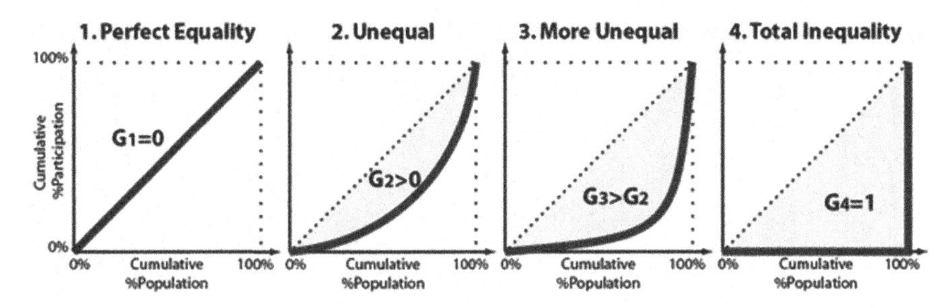

Fig. 2.13 Different Lorenz curves

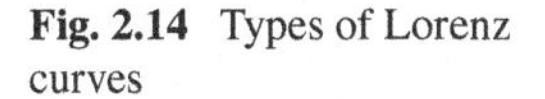
Fig. 2.14 Types of Lorenz curves

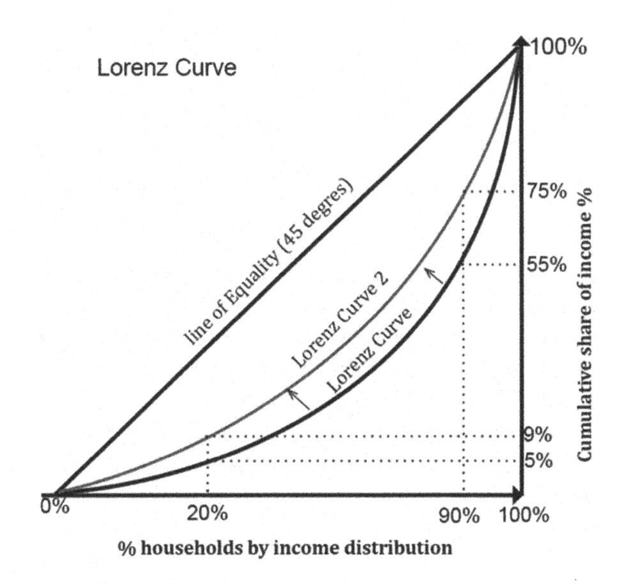

Lorenz curve: 20% of the total population contributed only 5% of the total income of the country and a predominant 90% of the population contributed just 55% of the total income. Hence, we can interpret that **the 10% of the affluent class (rich population) in the country contributed a significant 45% of total income!**

Lorenz curve 2: The second curve is slightly closer to the line of equality and shows a slight improvement in inequality when compared to the first curve. The low-income population in the country contributes a total of 9% of total income and 10% of the top affluent class in the country contributes a total of 25% of total income.

The Gini Coefficient

See Fig. 2.15.

This important statistical measure was discovered by the Italian statistician ***Corrado Gini in 1912.*** It's a very simple and interesting tool that describes levels of

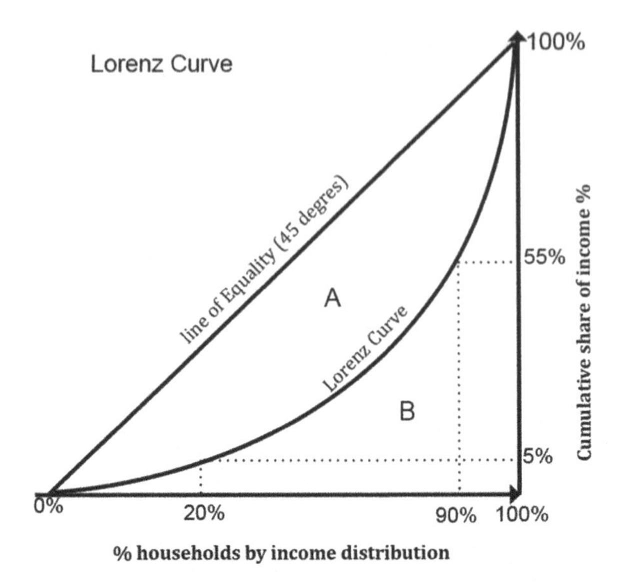

Fig. 2.15 Graph to explain Gini coefficient

inequality in terms of the area under the line of equality. With this, the Gini coefficient is defined as, Gini coefficient $= \frac{A}{A+B}$. The area under the line of equality: The area under the line of equality is a right-angle triangle, which is half of the area of a perfect square graph with an axis that ranges from 0 to 1.

Take a look at Fig. 2.16,

Area A: Area between the line of equality and the Lorenz curve. The higher the inequality, the higher the area of A. Gini coefficient is directly proportional to the value of A.

Fig. 2.16 Corrado Gini

Area B: Area under the line equality without Area A. This means everything else under the Lorenz curve is occupied by B.

- The Gini coefficient is not an absolute measurement of income.
- The Gini coefficient ranges from 0 to 1, where 0 depicts perfect equality and 1 depicts inequality.
- Both high- and low-income countries can have the same Gini coefficient, and it only talks of the dispersion of data.

A note on population and sample values

In statistics, **population** refers to a group of all units/observations under study.

- Sometimes, it becomes very difficult to study all the entities to draw conclusions, because to study a large set, it's extremely time-consuming and expensive too.
- There are also situations where the entities can be destroyed during such tests too. For example, if we need to check the quality of matchsticks produced by the factory, they have to be lit and so they lose their life. So the matchsticks need to be selected at random in a systematic manner to draw inferences.
- Sometimes it simply doesn't make sense to consider these many units, when it's easier to study and judge a random sample from the population.

Therefore, a **sample** refers to a few entities that are drawn from a population in a systematic meaningful way. Later, in this book, we will learn more about sampling and its techniques. For now, let's look at some terminologies that are used to describe the statistical values of a sample and population.

The statistical parameters that define a sample are called "**Sample statistic**", and those describing a population is called "**Population parameter**".

Figure 2.17 shows the different notations and their meaning regarding a sample or a population value. All these terminologies will be explained as and when required in this book. We have come across $\boldsymbol{n}$ for several samples in discrete and raw data examples, whereas for continuous data sets, we write $\boldsymbol{N}$ for the total number of units which is the summation of all the frequencies $\boldsymbol{f}$.

Sample Statistic		Population Parameters
$\bar{x}$	Mean	μ
S^2	Variance	σ^2
n	Number of units	N
p	Proportions	P
S	Satnadard Deviations	σ

Fig. 2.17 Sample statistic and population parameters

References and Research Articles

1. **Journal of Vocational and Technical Education, Volume 12, Number 2 (1996). Title**: Marketing and Marketing Effectiveness Perceptions of Ohio Vocational Education Planning District (Vepd) Superintendents. **Authors:** Deborah B. Catri, Director, Vocational Instructional Materials Laboratory, the Ohio State University and R. Kirby Barrick, Professor and Head, Department of Agricultural Education, The Ohio State University. Illustrates the use of means, SDs, ranges, and grouped frequency distributions to summarize data. URL: http://scholar.lib.vt.edu/ejournals/JVTE/v12n2/catri.html
2. **Medicine on-line—Current Medical Research and Opinion (1995), 13, No. 6, 299. Title**: Comparison of azelastine nasal spray and oral ebastine in treating seasonal allergic rhinitis. **Authors:** D. J. Conde Hernández, J. L. Palma Aqilar and J. Delgado Romero. Illustrate use of the mean, median, and range. URL: http://www.priory.com/cmro/cmro1091.htm
3. **Psychiatry On-Line. Title**: Therapeutic Group Work with Adolescent Refugees in the Context of Way and Its Stresses. **Authors:** Natasa Ljubomirovic MD. For each of the 8 variables, means

and standard deviations are presented for females, for males, and both groups combined. URL: http://www.priory.com/psych/refugee.htm
4. **Clinical and Diagnostic Laboratory Immunology. 2002 November; 9(6): 1235–1239. Title**: Use of Coefficient of Variation in Assessing Variability of Quantitative Assays. **Authors:** George F. Reed, * Freyja Lynn, and Bruce D. Meade. URL: http://www.ncbi.nlm.nih.gov/pmc/articles/PMC130103

Practice Data Sets—Coding

All the information below and many more links can be found on our GitHub Page: https://github.com/ElementaryStatisticalMethods/Book1.git

For Python Users: In Jupyter NoteBooks

1. **Practice data sets to understand quartiles and deviations**: master/Descriptive_Statstics_using_Python/Measure_of_Spread.ipynb.
2. **Comprehensive data exploration**. https://www.kaggle.com/pmarcelino/comprehensive-data-exploration-with-python
3. **Exploring the US College Scorecard Data understanding the central tendencies and partition of data**. https://www.kaggle.com/benhamner/exploring-the-us-college-scorecard-data
4. **Dispersion of data used to predict fraud in financial services**. https://www.kaggle.com/arjunjoshua/predicting-fraud-in-financial-payment-services

For R Programming Users

1. **Quantiles and its 6 uses in R programming**. https://statisticsglobe.com/quantile-function-in-r-example
2. **The avocado data set**. https://michael-franke.github.io/intro-data-analysis/Chap-02-03-summary-statistics-1D.html

Other References

1. **Comic on the arithmetic mean**: www.glassbergen.com, randy@glassbergen.com
2. **Median of Skewed Data**: https://en.wikipedia.org/wiki/Skewness
3. **Case study on median**: Seglen, P.O. 1992. The skewness of science. Journal of the American Society for Information Science 43(9):628-638
4. **Relationship between AM, HM and GM**: https://math.stackexchange.com/questions/2930098/why-are-am-gm-inequality-related-exercises-so-popular/2930146
5. **Partition Values**: https://axibase.com/use-cases/workshop/percentiles.html
6. **SAT Scores for percentiles**: Data from the National Centre for Educational Statistics
7. **Example for quartile deviation**. Women having their first child at various ages: Adapted from: Krantz (1992, p 190) The numbers do not sum to 100% owing to being rounded off.

8. The famous 3 sigma rule of the normal distribution: https://stats.stackexchange.com/questions/476677/understanding-standard-deviation-in-normal-distribution.
9. **Case study on year and rainfall**: Adapted from: Amy Goldwitz, the Davis Enterprise, March 4, 1998, pA1.
10. **Lorenz curve**: www.economicshelp.org
11. **Types of Kurtosis**: 2005, haron E Robinson Kurpius, Mary E, Stanford and Jason Love.

Chapter 3
Probability and Distributions

WHAT

Probability, Random variable, Theoretical Distributions and Normal distribution in detail.

WHY

- To deal with uncertain data using probability concepts.
- To understand theoretical distributions.

HOW

- By understanding the basic idea and uses of common distributions.
- Solving problems on various scenarios using probability concepts.

WHEN

- Dealing with the randomness of data.
- Deriving conclusions with reasonable estimates.
- Working with sample and population data.

S. Prasad, *Elementary Statistical Methods*,
https://doi.org/10.1007/978-981-19-0596-4_3

CHAPTER 3

- Understanding the concept of Probability and Random Variables.
- Bayes theorem.
- Theoretical distributions.
- Normal distribution.

WITH MR.STAT

WITH MISS TICS

- Case study on Psychology deceit.
- Probabilities from lie detectors.
- Conditional probability concepts using IRIS data set in R.
- Application of Bayes' theorem on Sudden Infant Death Syndrome (S.I.D.S)

3.1 Introduction: Probability and Uncertainty

"I might go home early", "It might rain today", "The chances of us bagging the contract is very less", "What are the odds of the patient surviving for another six months", "What is the possibility that we may get a seat in the next flight".

These kinds of sentences are commonly used in real life. Probability or possibility or chances of occurrence are used in those cases where there is uncertainty. If we throw a ball from a height and there is nothing to stop the ball midway, it will surely fall to the ground. There is no uncertainty here. But think of a dart thrown at a board. It might strike anywhere. A seasoned player is more likely to hit the target as compared to an unseasoned player. In this case, there is a probability or possibility whereas in the first case, there is only certainty.

Probability is thus a concept used for situations which have no certain outcomes. These could range from tossing a coin, reaching a destination on time, passing an examination, getting a job, writing a code correctly and so on.

3.2 Some Basic Terminologies

1. **EXPERIMENT**: Any situation which has an outcome is an experiment, for example, the tossing of a coin or buying a chocolate from a store (the brand of chocolate you buy may be based on the display counter). Based on their outcomes, experiments can be deterministic or random. An experiment which has a fixed or known outcome is deterministic. For example anything which is thrown will fall, you always buy the same brand of soap. These have no uncertainty and so are deterministic.

   If we do not know the outcome of the experiment but we are aware of all possible outcomes of the experiment, then this will be a random experiment; e.g. a coin is being tossed, we do not know the outcome till the toss actually occurs but we do know that the possible outcomes are head or tail, assuming that the coin does not roll away to a corner and we are not able to see what is the outcome.

Random experiments are also called as statistical experiments. Another commonly used word is probability experiment. Trials are conducted a large number of times to measure the chance of an event occurring in future. Statistical experiment or statistical observation can thus be defined as any random activity that results in a definite outcome (Fig. 3.1).

2. **EVENT**: An **event** is an outcome of a statistical experiment. For example, in a coin toss experiment, there are two events—head and tail. An event can be simple; that is, there is only one outcome, say $A = \{5\}$. If it has more than 1 outcome, then it is known as a compound event. For example $B = \{$Even numbers between 0 and 10, both inclusive$\}$ is compound as it has many outcomes associated with it, $B = \{2, 4, 6, 8, 10\}$.

We can define many events in Probability, some of them are as below:

a. **Equally Likely Events**: Two or more events are said to be equally likely if both events have an equal chance of occurrence. For example, when we toss an unbiased coin, heads and tails are equally likely to appear.
b. **Mutually Exclusive Events**: Two events A and B of a random experiment are said to be mutually exclusive if they cannot occur simultaneously. For example, when we roll an unbiased dice, we cannot get both 5 and 3 facing us. We have a chance of either 5 or 3.

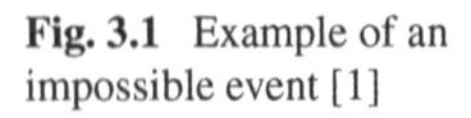

Fig. 3.1 Example of an impossible event [1]

c. **Impossible Events**: Events that are logically impossible to occur while experimenting are called impossible events. For example, getting the number 8 on an unbiased dice is certainly impossible. As shown in the illustration, getting balls from a bag of screws is an example of an impossible event.
d. **Certain/Sure Events**: These are events that are sure to happen when conducting the trials/experiment. "Head or tail" is a certain event connected with tossing an unbiased coin. The probability values of certain events are equal to 1.
e. **Independent Events**: Two events are independent if the occurrence or non-occurrence of one does not affect the occurrence or non-occurrence of the other. For example, two events such as a political party winning elections and cooking dal and rice for dinner are totally independent of each other.
f. **Dependent Events**: Two events are dependent if the occurrence or non-occurrence of one depends on the occurrence or non-occurrence of the other. Committing a crime and being caught by the police are two dependent variables.
g. **Null Events**: An event which has no outcome is called a null event. For example, choosing a green ball from a box of blue balls is a null event. It is denoted by $\varphi = \{\}$. The probability of a null event is always 0. Some authors refer to a null event as a set of impossible events,
h. **Favourable Events**: Events which result in the occurrence of any event are called favourable to that event. For example, Let A = event of obtaining an even number between 0 and 10. Then the set $\{2, 4, 6, 8, 10\}$ are favourable to event A. Similarly, $A' = \{1, 3, 5, 7, 9\}$ is the non-favourable event to A.
i. **Complementary Events**: Events that complete the sample space are called complementary events. They are denoted by A' or A^c.

   E.g.: Let S = {Blue, green, yellow, pink} If A = {Blue, green} then A' = {yellow, pink} Note that complimentary events are mutually exclusive but mutually exclusive events need not be complimentary.

For example, $C = \{\text{yellow}\}$ and $D = \{\text{Blue}]$ are mutually exclusive but they are not complimentary events.

3. **Sample Space** can be defined as a set of all possible simple events of an experiment. For example, sample space for tossing a coin = {Head, Tail}. All these events are listed within a pair of parentheses "{}". Sample space can be finite or infinite. Finite sample space is the one which has a finite number of events in it. For example, tossing a dice once will result in the sample space $S = \{1, 2, 3, 4, 5, 6\}$.

An infinite sample space occurs when there are an infinite number of possible outcomes of a random experiment. They are of two types, depending on the situation.

a. Countably infinite and
b. Uncountably infinite.

When we are able to arrange all outcomes of an experiment in a series and count them, we will get a countably infinite sample space. A few examples of such sample spaces are:

a. Mistakes made by a typist on a page. Theoretically, they can be 0,1, 2 .... infinity errors.
b. Let a person be asked to toss a coin till he/she gets a head. Then X be the number of tosses before getting head. The X can be 0 if the head occurs on the first toss. X can be 1, 2 and so on.

When the number of possible outcomes is very large and it is not possible to count them sequentially, we say that the sample space is uncountably infinite. A good example would be the set of real numbers. Any situation where there are many possible values, and it isn't possible to enumerate will be an example of such sample space.

Examples of such sample spaces are:

1. Weight of a grown person (X), which can be anything from 45 to 90 kg. There are many possible values of X like 45.123 kg, 67.34 kg and so on.
2. Waiting for an event will also give rise to such a sample space. For example X can be anything from a second to many seconds/minutes. Like 0.02 s, 56.68 s and so on (Fig. 3.2).

3.3 Probability Definitions

1. **Mathematical (or Classical or "apriori") Probability** [2]

In a random experiment, there are **"n"** exhaustive, mutually exclusive and equally likely outcomes, out of which **"m"** is favourable to the occurrence of an event E. Then the probability of an event E, usually denoted by $P(E)$, is given by:

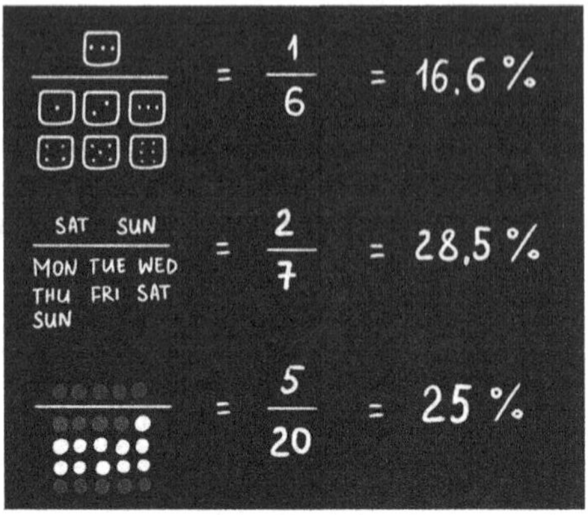

Fig. 3.2 Classical definition of probability

$$P(E) = \frac{\text{Number of favourable cases}}{\text{Total number of exhaustive cases}} = \frac{m}{n}$$

2. **Statistical (or Empirical) Probability**

If an experiment is performed repeatedly under essentially homogeneous and identical conditions, then the limiting value of the ratio of the number of times the event occurs to the number of trials, as the number of trials becomes indefinitely large, is called the probability of happening of the event, it is assumed that the limit is finite and unique. If "*n*" denotes the number of trials and "**m**" denotes the number of times event E has occurred, the prability of the happening of E is given by:

$$P(E) = \lim_{n \to \infty} \frac{m}{n}$$

3. **Axiomatic Probability**

Probability can be defined as a set function *P*(*E*) which assigns to every event *E*, a number, known as the **"Probability of E"** such that,

1. The probability of an event *P*(*E*) is greater than or equal to zero $P(E) \geq 0$.
2. The probability of the sample space is equal to one $P(\Omega) = 1$.

3.4 Summary of Basic Probability Rules

The notation *P*(*A*) designates the probability of event *A*.

1. **P (entire sample space) = 1**
2. For any event **A:** $0 \leq P(E) \leq 1$
3. If A^c designates the complement of *A*, then $P(A^c) = 1 - P(A)$
4. Events *A* and *B* are independent events if *P*(*A*) = *P*(*A*, given *B*).
5. **Multiplication rules**:

General: $P(A \text{ and } B) = P(A) * P(B, \text{given } A)$
Independent events: $P(A \text{ and } B) = P(A) * P(B)$

6. **Conditional Probability**: $P(A \text{ given } B) = \frac{P(A \text{ and } B)}{P(B)}$
7. Events A and B are mutually exclusive if **P (A and B) = 0**
8. **Addition rules**:

General events: $P(A \text{ or } B) = P(A) + P(B) - P(A \text{ and } B)$
Mutually exclusive events: $P(A \text{ or } B) = P(A) + P(B)$.

Probability assignments

1. A probability assignment is based on intuition and incorporates past experience, judgement, or opinion to estimate the likelihood of an event.
2. A probability assignment based on **relative frequency** uses the formula below: $\text{Probabilityofanevent} = \frac{f}{n}$, where **f** is the frequency of the event that occurred in a sample of **n** observations.
3. A probability assignment based on equally likely outcomes uses the formula,

$$\text{Probability of an event} = \frac{\text{Total number of favourable outcomes to an event}}{\text{Total number of outcomes}}$$

Law of Large Numbers: In the long run, as the sample size increases, the relative frequencies of outcomes get closer and closer to the theoretical (or actual) probability value.

Multiplication rule for independent events:
$P(A \text{ and } B) = P(A) * P(B)$

Addition rule of mutually exclusive events A and B:

$P(A \text{ or } B) = P(A) + P(B)$

Case study: A magazine gave the information shown in the table about ages of children receiving toys. The percentages represent all toys sold (Table 3.1).

What is the probability that the toy is purchased for someone?

Table 3.1 Table of age and percentage of toys bought

Age (years)	Percentage of toys (%)
2 and under	16
3–5	20
6–9	28
10–12	13
13 and over	23

(a) 6 years and older?
(b) 12 years old and younger?
(c) Between 6 and 12 years old?
(d) Between 3 and 9 years old?

A child between 10 and 12 years old looks at this probability distribution and asks, "Why are people more likely to buy toys for kids older than I am, who are aged 13 and above, than for kids in my age group which is between 10 and 12"? How would you respond?

Suitable Tool: Addition rule for mutually exclusive events.

Solution:

(a) P (6 years old or older)

$= P\ (6\text{–}9) + P\ (10\text{–}12) + P\ (13 \text{ and over})$
$= 0.28{+}0.13{+}0.23{=}\ \mathbf{0.64}$

(b) P (12 years or younger)

$= P\ (2 \text{ and under}) + P\ (3\text{–}5) + P\ (6\text{–}9) + P\ (10\text{–}12)$
$= 0.16 + 0.20 + 0.28 + 0.13 = \mathbf{0.77}$

(c) P (Between 6 and 12)

$= P(6\text{–}9) + P(10\text{–}12)$
$= 0.28 + 0.13 = \mathbf{0.41}$

(d) P (Between 3 and 9)

$= P(3\text{–}5) + P(6\text{–}9)$
$= 0.20 + 0.28 = \mathbf{0.48}$

Reason: The keyword here is **"or"**. A child can belong to only one age group at a time. So, a child who is aged 6 years is mutually excluded from the other age groups except for the age group 6–9. The category 13 and over contains far more ages than the group 10–12. Therefore, it is not surprising that more toys are purchased for this group since this group contains more children.

Converting odds in favour and odds in against to probability

A brief explanation of calculating probability using odds in favour of and odds against the event of study. If "odds of", "odds for" and "odds in favour of" any event *A* are

m:n, then the number of favourable outcomes for A is m and the number of outcomes not favourable to A is n, that is, the total number of trials $= m + n$. So, $P(A) = m/m + n$.

Similarly, if we have odds against A are m_1:n_1, then $P(A) = n/m + n$.

$$\text{Odds in favour} = \frac{\text{Number of favourable choices}}{\text{Number of unfavourable choices}}$$

$$\text{Odds against} = \frac{\text{Number of unfavourable choices}}{\text{Number of favourable choices.}}$$

Example 1 If odds in favour of X solving a problem are 5–3 and odds against Y solving the same problem are 1–6. Find the probability for:

(i) X solving the problem.
(ii) Y solving the problem.

Solution

Given odds in favour of X solving a problem are 5 to 3.

The number of favourable outcomes = 5
The number of unfavourable outcomes = 3
Total number of events = 5 + 3 = 8

(i) **X solving the problem**
P(X) = P (solving the problem) = 5/(5 + 3) = 5/8
Given odds against Y solving the problem are 1 to 6
Number of favourable outcomes = 6
Number of unfavourable outcomes = 1

(ii) **Y solving the problem**

P(Y) = P (solving the problem) = 6/(1 + 6) = 6/7

Conditional Probability

The conditional probability of event B is the probability that the event will occur given that event A has already occurred. This probability is written as **P (B|A)** and read as **the probability of B given A.** In this case, the sample space reduces.

Case 1: In the case where events A and B are independent, then the conditional probability of event B given that event A is simply the probability of event B, that is $P(B)$.

Case 2: If events A and B are not independent, then the probability of the intersection of A and B (the probability that both events occur) is defined by.

P (A and B) = P (A)* P (B|A)

From this definition, the conditional probability P(B|A) is easily obtained by dividing the P(A and B) by P(A):

$$\text{P(B|A)} = \frac{\text{P (A and B)}}{\text{P (A)}}$$

General multiplication rule for any events:

P (A and B) = P (A)* P (B, given that A has occurred)

P (A and B) = P (B)* P (A, given that B has occurred)

If two events cannot possibly occur together, they are said to be mutually exclusive or disjoint. In particular, events A and B are mutually exclusive if P(A and B) = 0.

Case Study: Salary raise: Does it pay when asked for a raise? A national survey of heads of households showed the percentage of those who asked for a raise and the percentage who got one (USA Today). According to the survey, of women interviewed, 26% asked for a raise, and of those women who have asked for a raise, 42% received the raise. If a woman is selected at random from the survey population of women, find the following probabilities of:

1. P (woman asked for a raise).
2. P (woman received raise, given she asked for one).
3. P (woman asked for a raise and received raise).

Suitable Tool: Multiplication rule of probability.

Reason: When one needs to find the probability of the occurrence of simultaneous events, the multiplication rule is used. The keyword here is "and".

Solution:

1. P (asked) = 0.26 or 26%.
2. P (received, given asked) = 0.42 or 42%.
3. P (ask and receive) = (0.26) (0.42) = **0.1092 or 10.92%.**

Case Study [4]

A lie detector test in the book "*Chances: Risk and Odds in Everyday Life*" by James Burke says that there is a 72% chance a polygraph test (lie detector test) will catch a person who is indeed lying. Furthermore, there is approximately a 7% chance that the polygraph will falsely accuse someone of lying.

a. Suppose that a person answers 90% of a long battery of questions truthfully. What percentage of the answers will the polygraph wrongly indicate are lies?
b. Suppose that a person answers 10% of a long battery of questions with lies. What percentage of the answers will the polygraph correctly indicate are lies?
c. Repeat parts (a) and (b) if 50% of the questions are answered truthfully and 50% are answered with lies.
d. Repeat parts (a) and (b) if 15% of the questions are answered truthfully and the rest are answered with lies.

Suitable Tool: Conditional Probability

Solution: *P* (report lie, given person is lying) = 0.72 and P (report lie, given person is not lying) = 0.07.

a. P (person is not lying) = 0.90;

= P (person is not lying and polygraph reports lie)
= P (person is not lying) * P (reports lie, given person is not lying)
= (0.90) * (0.07) = **0.063 or 6.3%**

b. P (person is lying) = 0.10

= P (person is lying and polygraph reports lie)
= P (person is lying) × P (reports lie, given person is lying)
= (0.10) * (0.72) = **0.072 or 7.2%**

c. P (person is not lying) = 0.5

= P (person is lying) = 0.5
= P (person is not lying and polygraph reports lie)
= P (person is not lying) × P (reports lie, given person is not lying)
= (0.50) * (0.07) = **0.035 or 3.5%.**

d. P (person is not lying) = 0.15

= P (person is lying) = 0.85
= P (person is not lying and polygraph reports lie)
= P (person is not lying) × P (reports lie, given person is not lying)
= (0.15) * (0.07) = **0.0105 or 1.05%**.
= P (person is lying and polygraph reports lie)
= P (person is lying) × P (reports lie, given person is lying)
= (0.85) * (0.72) =**0.612 or 61.2%.**

Permutations and Combinations

Many complex probability problems can be solved by simply understanding the concepts of permutations and combinations. In the concept of combinations, there can be ***n*** ways of selecting **m** units. Permutations are also similar but the order in which the units are chosen matters. As shown in Figs. 3.3 and 3.4.

Counting rule for permutations: The number of ways to arrange ***n*** distinct objects in order taking them ***r*** at a time is $\mathbf{{}^{n}P_{r}} = \frac{\mathbf{n!}}{\mathbf{(n-r)!}}$ where **n** and **r** are whole numbers and $n \geq r$.

Counting rule for combinations: The number of combinations of n objects taken r at a time is $\mathbf{{}^{n}C_{r}} = \frac{\mathbf{n!}}{\mathbf{r!(n-r)!}}$ where n and r are whole numbers and $n \geq r$.

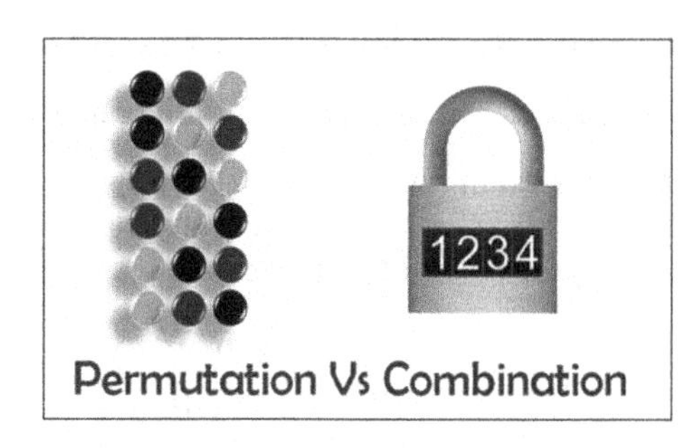

Fig. 3.3 Permutations and combinations

Fig. 3.4 Thomas Bayes

How to determine the number of outcomes of an experiment

1. Use the multiplication rule or tree diagram, if the experiment consists of a series of stages with various outcomes.
2. Use the permutations rule, ${}^{n}P_{\mathrm{r}} = \frac{n!}{(n-r)!}$ if the outcomes consist of ordered subgroups of r items taken from a group of n items.
3. Use the combinations rule ${}^{n}C_{\mathrm{r}} = \frac{n!}{r!(n-r)!}$ if the outcomes consist of non-ordered subgroups of r items taken from a group of n items.

The very famous Bayes Theorem

This theorem is vital and its applications are omnipresent in almost all fields of study. This is because it establishes a mathematical basis for probabilistic inference. It was named after eighteenth century British mathematician Thomas Bayes.

3.5 Bayesian Basics

Prior: Probability distribution representing knowledge or uncertainty of a data object before (prior) observing it.

Posterior: Conditional probability distribution representing what parameters are likely after (posterior) observing the data object.

Posterior distribution = Prior * Likelihood

Likelihood: The probability of falling under a specific category or class.

The Bayesian Statement: If $E_1, E_2 \ldots E_n$, are mutually disjoint events with, $P(E_i) \neq 0$, $(i = 1, 2, 3, \ldots, n)$, then for any arbitrary event A which is a subset of $\bigcup_{\mathrm{i}}^{\mathrm{n}} \mathrm{E_i}$ such that $P(A) > 0$, we have

$$P(E_i|A) = \frac{P(E_i) * P(A|E_i)}{\sum_{i=1}^{n} P(E_i) * P(A|E_i)} = \frac{P(E_i) * P(A|E_i)}{P(A)} i = 1, 2 \ldots, n$$

When to Use?

(i) The sample space is partitioned into a set of mutually exclusive events $\{A_1, A_2, ..., A_n\}$.
(ii) Within the sample space, there exists an event B, for which $P(B) > 0$.
(iii) If there is prior knowledge of a situation, it is required to predict the probabilities of any outcome.

An interesting thought: Richard Dawkins, a professor of the University of Oxford, argues in his book "The God Delusion", against the use of Bayes's theorem for assigning a probability to God's existence. Using Bayes's theorem, if we calculate the probability of God-given our experiences in the world (the existence of evil, religious experiences, etc.) and assign numbers to the likelihood of these facts given existence or nonexistence of God, as well as to the prior belief of God's existence, i.e. the probability we would assign to the existence of God if we had no data from our experiences. Dawkins's argument is with the lack of data put into this formula by those employing it to argue for the existence of God. The equation is perfectly accurate, but the numbers inserted are, to quote Dawkins, "not measured quantities but and personal judgements, turned into numbers for the sake of the exercise".

Another interesting article to be found in PMC-US National Library of Medicine & National Institutes of Health titled "Bayes' Theorem to estimate population prevalence from Alcohol Use Disorders Identification Test (AUDIT) scores" by David R Foxcroft, Kypros Kypri, and Vanessa Simonite uses Bayes' Theorem to estimate the difference in the prevalence of a disorder in two groups whose test scores are obtained.

Case Study: Suppose that a woman in her forties goes for a mammogram and receives bad news: a "positive" mammogram. However, since not every positive result is real, what is the probability that she has breast cancer? Given that the fraction of women in their forties who have breast cancer is 0.014 and the probability that a woman who has breast cancer will get a positive result on a mammogram is 0.75. The probability that a woman who does not have breast cancer will get a false positive on a mammogram is 0.1 (Adapted from The New Yorker, Jan 25, 2013).

Suitable Tool: Bayes' theorem.

Reason: We have prior knowledge of all possible probabilities.

Discussion: To calculate this, we proceed as follows:
See Fig. 3.5

(i) The fraction of women in their forties who have breast cancer is **0.014**, which is about one in seventy. The fraction who do not have breast cancer is therefore $1 - 0.014 = \mathbf{0.986}$. These fractions are known as the prior probabilities.
(ii) The probability that a woman who has breast cancer will get a positive result on a mammogram is 0.75. The probability that a woman who does not have breast cancer will get a false positive on a mammogram is 0.1. These are known as conditional probabilities.

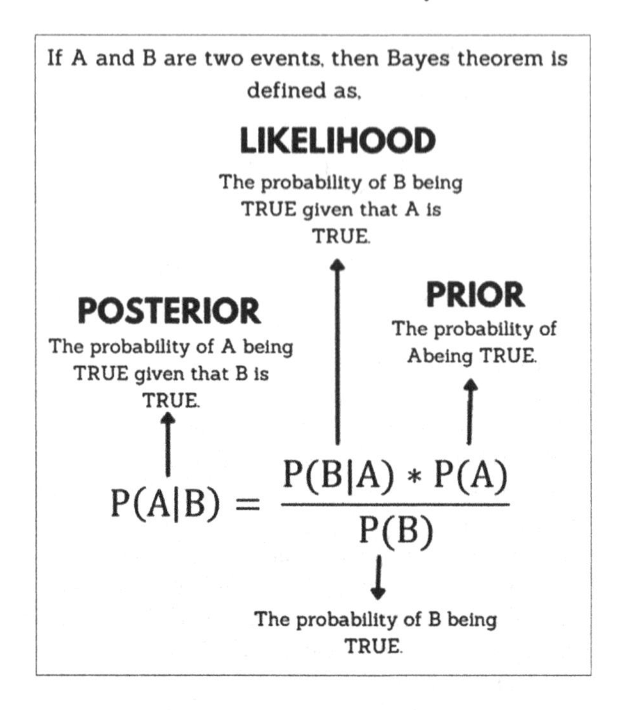

Fig. 3.5 Explanation of formula—Bayes' Theorem

(iii) Applying Bayes' theorem, we can conclude that, among women who get a positive result, the fraction who have breast cancer is

$\frac{0.014X0.75}{(0.014X0.75)+(0.986X0.1)} = 0.1$, Approximately. That is, Once We Have Seen the Test Result, the Chance is About Ninety Per Cent that It is a False Positive.

Case Study [11]

One famous case of a failure to apply Bayes' theorem involves a British woman, Sally Clark. After two of her children died of sudden infant death syndrome (SIDS), she was arrested and charged with murdering her children. Pediatrician Roy Meadow testified that the chances that both children died of SIDS were 1 in 73 million. He got this number by squaring the odds of one child dying of SIDS in similar circumstances (1 in 8500).

Because of this testimony, Sally Clark was convicted. The Royal Statistics Society issued a public statement decrying this "misuse of statistics in court", but Sally's first appeal was rejected. She was released after nearly 4 years in a women's prison where everyone else thought she had murdered her children. She never recovered from her experience, developed an alcohol dependency and died of alcohol poisoning in 2007.

The statistical error made by Roy Meadow was, among other things, to fail to consider the *prior probability* that Sally Clark had murdered her children. While two sudden infant deaths may be rare, a mother murdering her two children is even rarer.

What tool is suitable in this situation? We need to find the conditional probabilities of the various possible events causing death, *given* the fact that the children died.

If H is some hypothesis, for example, that both of Sally Clark's children died of cot death—and D is some data, that both children are dead we want to find the probability of the hypothesis given the data, which is written *P(H|D)*.

Let A be for the alternate hypothesis - that the children did not die of cot death.

Discounting all other possibilities, for example, that someone else murdered both children, or that Sally Clark murdered only one of them, or that they died of natural causes other than cot death.

$$P(H|D) = \frac{P(D|H) * P(H)}{P(D|H) * P(H) + P(D|A) * P(A)}$$

P(D/H) = 1 as it is the probability that two of the children are dead, given that two of the children have died of natural causes. *P(H)* = 1/1,00,000. (1/73 million can be approximated to this figure).

P(D/A) is the probability that the children died given that they did not die of natural causes. In other words, it is the probability that a randomly chosen pair of siblings will both be murdered. This is the most difficult figure to estimate. Statistics on such double murders are difficult to get, because child murders are so rare (far, far rarer than cot deaths) and because in most cases, someone known to have murdered once is not free to murder again. So, we take the Home Office statistic that fewer than 30 children are known to be murdered by their mother each year in England and Wales. Since 6,50,000 are born each year, and murders of pairs of siblings are rarer than single murders, we should use a value much smaller than 30/6,50,000 = 0.000046.

So, we can consider a number ten times as small here, i.e. 0.000046 * 10 = 0.0000046. Thus,

$$\begin{aligned} P(H|D) &= \frac{P(H)}{P(H) + P(D|A) * (1 - P(H))} \\ &= \frac{0.0001}{0.0001 + 0.00046 * (1 - 0.0001)} > \frac{2}{3} \end{aligned}$$

This is the probability that Sally Clark is innocent. Thus, the Bayes theorem gives a correct picture.

Case Study [5]

We want to know how a change in interest rates would affect the value of a stock market index. All major stock market indexes have a plethora of historical data available so you should have no problem finding the outcomes for these events with a little bit of research. For our example, we will use the data below to find out how a stock market index will react to a rise in interest rates (Table 3.2).

Suitable Tool: Bayes' theorem

Table 3.2 Stock prices and interest rates

Stock price	Interest rates			
		Decline	Increase	Unit frequency
	Decline	200	950	1150
	Increase	800	50	850
	Total	1000	1000	2000

Reason: We have prior probabilities and we need to update them with new information available.

P(SI) = The probability of the stock index increasing.

P(SD) = The probability of the stock index decreasing.

P(ID) = The probability of interest rates decreasing.

P(II) = The probability of interest rates increasing.

P(SD) = 1150/2000 -prior probability.

What we find after applying Bayes' theorem is a posterior probability.

$$P(SD|II) = \frac{P(SD) * P(II|SD)}{P(II)} = \frac{0.575 * 0.826}{0.5}$$

$$= \frac{0.47495}{0.5} = 0.9499 \approx 95\%$$

Thus, we can use the outcomes of historical data to base our beliefs on and from which we can derive newly updated probabilities.

Random Variable

A random variable is a function that assigns a real number to each point in a sample space **S**. They may be categorized as discrete variables with precise values or continuous variables that can take any values within an interval. A random variable is denoted by capital letters such as X and Y. A random variable is a process of assigning values to different cases.

Example 2 Let S be a sample space of various states. Then we can define a random variable

$X = 1$, if it has achieved the literacy target.

$X = 0$, if it has failed to achieve the target.

Example 3 Let X be the random variable denoting the height of a randomly selected child from a group of 8-year-old children. It can take various values according to the heights of children.

Example 4 Let's consider a simple and most common example used in probability, tossing an unbiased coin. We wish to determine the probability of getting 0, 1, 2 or 3 heads, simultaneously when 3 coins are tossed. The experiment revealed the following results (Table 3.3).

Table 3.3 Table showing results of tossing a coin

Outcome	First toss	Second toss	Third toss
1	Head	Head	Head
2	Head	Head	Tail
3	Head	Tail	Head
4	Head	Tail	Tail
5	Tail	Tail	Head
6	Tail	Tail	Tail
7	Tail	Head	Head
8	Head	Head	Tail
9	Tail	Head	Tail
10	Head	Head	Head

Let's calculate the probability of heads using the simple probability formula. Take a note of how X is defined and also the total probability of all possible outcomes is 1 (Table 3.4).

Probability distribution of a discrete random variable

- A random variable has a probability distribution whether it is discrete or continuous.
- A probability distribution is an assignment of probabilities to the specific values of a random variable or a range of values of the random variable.

Features of the probability distribution of a discrete random variable

1. The probability distribution has a probability assigned to each value of the random variable.
2. The sum of all assigned probabilities must be 1.

The mean and standard deviation of a discrete population probability distribution is obtained by using the following formulae: $\mu = \sum x * P(x)$. Here, μ (population mean) is called the **expected value of x (sample).** $\sigma = \sqrt{\sum P(x) * (x - \mu)^2}$ is called the standard deviation of x (underlying population standard deviation) where x is the value of the random variable, $P(x)$ is the probability of that variable, and the sum $\sum$ is taken for all the values of the random variable.

Table 3.4 Calculating the probability of heads

Number of heads (X)	Number of outcomes	Probability
0	1	1/10 = 0.1
1	3	3/10 = 0.3
2	4	4/10 = 0.4
3	2	2/10 = 0.2
Total	10	1

A Discrete Random Variable

A variable that can assume only a countable number of real values and for which the value of the variable depends on a chance is called a discrete random variable. In other words, a real-valued function defined on a discrete sample space is called a discrete random variable. Other names are the discrete stochastic variable or the discrete chance variable.

	Discrete	Continuous
Type	Countable	Not countable
Definition	Probability mass function	Probability density function
Distributions	Binomial and Poisson distribution	Gaussian and exponential distribution

Example 5 number of accidents in a month, number of telephone calls per unit time, number of successes in **"n"** trials.

A Continuous Random Variable

A random variable X is said to be continuous if it can take all possible values (integral as well as fractional) between certain defined limits. In other words, a random variable is said to be continuous when its different values cannot be put in 1–1 correspondence with a set of positive integers, for examples, the age, height and weight of adolescents, etc. (Fig. 3.6).

Example 6 Are the following discrete or continuous random variables?

Different scenarios

Scenario 1. Time is taken by a student to pay a fine when he has returned a book late to the library.

Variable: Continuous random variable.

Reason: Time is a continuous random variable as it can take any value.

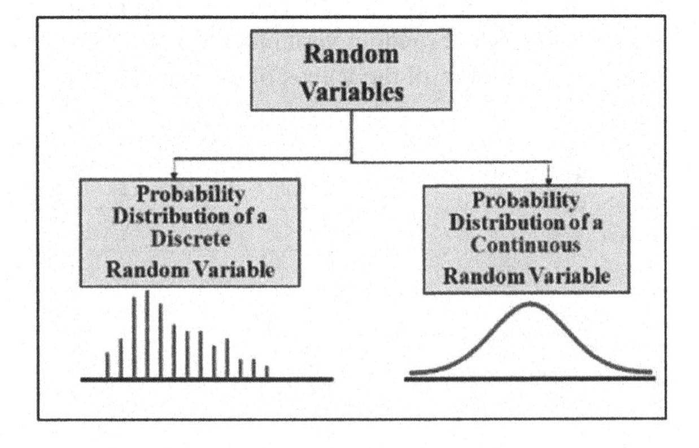

Fig. 3.6 Discrete and continuous random variable

Scenario 2. The number of cheques that bounced in a bank on a particular day.

Variable: Discrete random variable.

Reason: The number of cheques that bounced can be 0, 1, 2… and so on, and it is a discrete random variable.

Scenario 3. Number of people who voted in the recent election in a locality.

Variable: Discrete random variable.

Reason: This is a count, starting from 0, 1, 2… so it is discrete.

Scenario 4. Litres of petrol one can fill in a vehicle tank.

Variable: Continuous random variable.

Reason: This is regarding volume, which can take any value, so continuous.

Expected Value and Variance of a Random Variable

Generally, random variables can be characterized and dealt with effectively for practical purposes by considering quantities called their expectation. Mathematical expectation (expected value) is the average value of a random phenomenon.

The expected value of a **discrete random variable** X with **probability mass function (p.m.f) of** $f(x)$ is given by: $E(X) = \sum_{x} x * f(x)$, provided the right-hand series is absolutely convergent (Table 3.5).

The expected value of a **continuous random variable** X with probability density function (pdf) $f(x)$ is given by: $E(x) = \int_{-\infty}^{\infty} x * f(x).dx$, provided the right-hand integral is absolutely convergent. The variance of a random variable X is given by: $\mathbf{V(x) = E(x^2) - [e(x)]^2}$

Example 7 A private commercial marketing company wished to investigate whether the customer was influenced to buy the product by watching the TV advertisement.

Solution

We can treat the information shown as an estimate of the probability distribution because the events are mutually exclusive and the sum of the percentage is 100%. Let's figure out the mean/expected value and variance of the distribution (Table 3.6).

Example 8 An investor would always compare the returns and risk of several portfolios for understanding his profits. Investment type I is a balanced and guaranteed fund with low-risk, large-cap funds while type II investment predominantly involves

Table 3.5 Distribution and type of random variables

Number of people influenced	1	2	3	4	5
Percentage of buyers	31	30	25	9	5

Table 3.6 Mean and variance of the influence on buyers

Number of people influenced	Percentage of buyers (%)	Expected value	Variance
1	31	$1 * 0.31 = 0.31$	$1^2 * 0.31 = 0.31$
2	30	$2 * 0.30 = 0.6$	$2^2 * 0.6 = 2.4$
3	25	$3 * 0.25 = 0.75$	$3^2 * 0.75 = 6.75$
4	9	$4 * 0.09 = 0.36$	$4^2 * 0.36 = 5.76$
5	5	$5 * 0.05 = 0.25$	$5^2 * 0.25 = 6.25$

medium-risk mid-cap funds. To evaluate these funds' performance, an investor analyses them in 3 economic market conditions which are recession, stable economy and expanding economy with their subjective probabilities of occurrence. Let's look at the possible inference that can be drawn from the data for an investment of every 1000 rupees.

Solution

Take a look at the interesting numbers in the solution, the mean returns of both the investment types are equal to 117.5 whereas type II investment has a lesser standard deviation compared to type I. An investment that has a higher mean and lesser variation is considered the most desired investment. Even though the type of funds in type II investment are of medium risk, their variations in different market conditions are more stable than that of type I investment which predominantly consists of low-risk funds. Therefore, with the assumed probabilities, we can conclude that it's worth taking a higher risk and would suggest the investor go with type II investment than type I (Table 3.7).

Table 3.7 Economic market condition and investment type

Probability	Economic market condition	Investment	Investment
		Type I	Type II
0.35	Recession	−100	−200
0.5	Stable economy	+100	+50
0.15	Expanding economy	+250	+350

Theoretical Distributions

Definition

Theoretical distributions are distributions based on mathematical formulae and logic rather than empirical observations (raw scores). They are used in statistics to determine probabilities. Here we allocate probability to different frequency points/class intervals, out of the total frequency.

Decision-making becomes challenging when we deal with facts that are uncertain and random. This is where the concepts from probability distributions guide us. Probability distributions can be called logical distributions as these distributions combine some known facts and some reasonable estimates to build a mathematical model that results in some realistic solutions amidst uncertainty (Fig. 3.7 and Table 3.8).

Bernoulli Distribution

Jakob Bernoulli, A Swiss mathematician from the Bernoulli family was the founder of this distribution. He also introduced the first principles of calculus variations.

Bernoulli distribution is one of the simplest distributions in statistics where the random variable X can take just two values, which are 0 and 1.

Example 9 Let us consider the chances of waking up tomorrow without an alarm. If you wake up as expected, then $X = 1$ else if you continue to sleep $X = 0$. The probability of you waking up without an alarm is defined by **p**.

Table 3.8 Mean, variance and standard deviation of investments type

Probability	Economic condition	Investment		Mean		Standard deviation	
		Type I	Type II	Type I	Type II	Type I	Type II
0.15	Recession	−50	−200	−7.5	−30	50	77.46
0.5	Stable Economy	75	50	37.5	25	75	35.36
0.35	Expanding Economy	250	350	87.5	122.5	250	207.06
Total		275	200	117.5	117.5	375	319.88

Fig. 3.7 Jacob Bernoulli

Therefore, the parameter that defines the distribution is **p** which is the probability of $X = 1$. The value of $1 - p$ is the probability of $X = 0$ which is also denoted as **q**.

The p.m.f function is defined as: $\mathbf{p^x * (1 - p)^{1-x}}$

$$P(X = x) = \begin{cases} p \text{ for } x = 1 \\ 1 - p \text{ for } x = 0 \end{cases}$$

The mean and variance of the distribution are:

Mean [X] $= E[X] = \mathbf{p}$

Var [X] $= E[X^2] - (\mathrm{E[x]})^2 = p^* (1 - p) = \mathbf{pq}$

Some Key Points:

- In certain cases, p and q terms are also referred to as success and failure rates, respectively. Therefore, Bernoulli can be used only when there are two possible outcomes for an event.
- Both these events have a fixed probability of occurrence.
- **Independence of trials:** That means the probabilities must remain the same throughout the trials. Like in the tossing of a coin, p is always equal to 0.5
- Bernoulli and Binomial distribution (explained after Bernoulli in this chapter) are very closely related. **We can also say that the Binomial distribution is Bernoulli when n = 1.**
- Bernoulli distribution forms a strong base for many other distributions such as binomial, geometric distribution and negative binomial distribution.
- Use of Bernoulli in Epidemiology: While modelling clinical trials, drug effectiveness surveys, etc. We come across data with events related to death and disease. Bernoulli distribution answers the primary questions related to the probability of a person being infected with the disease or having died due to the disease. Based on this key parameter, further extensive research and analytics can easily be performed.

Example 10

- The chances that a random email you receive is spam or not spam.
- A cricketer either wins or loses a match.
- You may or may not accurately ace a dart game.
- Your boss is going to the office early tomorrow or not.

Example 11 The probability that you can clear the Actuarial Science entrance examination in the very first attempt is 0.8. Model this information using Bernoulli distribution. Let p be the probability of clearing the entrance examination which is given as 0.8

Therefore, the p.m.f is,

When $x = 1$, $0.8^1 * 0.2^0 = \mathbf{0.8}$

When $x = 0$, $0.8^0 * 0.2^1 = \mathbf{0.2}$

$E(x) = p = \mathbf{0.8}$

$\text{Var}(x) = p * q = 0.88 * 0.2 = \mathbf{0.16}$

Binomial Distribution

The prefix "bi" refers to two or twice. Even in this distribution, there are two possible outcomes and it is a continuation of Bernoulli with many trials. Which means $n > 1$. A discrete random variable X is said to follow a binomial distribution if its probability mass function (p.m.f) is of the form,

$$p(x) = P[X = x] = \begin{cases} n_x^c p^x q^{(n-x)}, x = 0, 1, 2, \ldots, n \text{ and } 0 < p < 1 \text{ and } p + q = 1 \\ 0, \text{ otherwise} \end{cases}$$

Example 12 Binomial models are used to calculate the

- The number of people violating traffic rules in a day.
- The number of times a lake overflows due to excess rain.
- The number of legal cases which face delayed justice due to time issues.
- The number of times the AC is repaired in hotels in summer, etc.

Characteristics

- The number of trials ***n*** is finite.
- The trials are dichotomous.
- The trials are independent.
- The probability of success ***p*** remains the same for all trials.

Mean $= \mu = E(x) = \text{np}$.
Variance $= \sigma^2 = \text{np}(1 - \text{p})$
Standard deviation $= \sigma = \sqrt{\text{np}(1 - \text{p})}$
where $n =$ Number of trials, $p =$ probability of success, $1 - p =$ probability of failure.

Case Study 1: Psychology Deceit [6]
Aldrich Ames is a convicted traitor who leaked American secrets to a foreign power. Yet Ames took routine lie detector tests and each time passed them. How can this be done? Recognizing control questions, employing usual breathing patterns, biting one's tongue at the right time, pressing one's toes hard on the floor, and counting backwards by 7 are countermeasures that are difficult to detect but can change the results of a polygraph examination (Adapted from Lies!!! Lies!! Lies!!! The Psychology of Deceit, by C.V. Ford, professor of psychiatry, University of Alabama). In fact, it is reported in Professor Ford's book that after 20 min of instruction by "Buzz" Fay (a prison inmate), **85% of those trained were able to pass the polygraph examination even when guilty of a crime**. Suppose that a random sample of nine students (in a psychology laboratory) are told a "secret" and then given instructions on how to pass the polygraph examination without revealing their knowledge of the secret. What is the probability that,

a. Are all the students able to pass the polygraph examination?
b. More than half the students can pass the polygraph examination?
c. No more than four of the students can pass the polygraph examination?
d. Do all the students fail the polygraph examination?

Suitable Tool: Binomial distribution

Reason: The number of trials $\boldsymbol{n} = 9$ is finite

- The trials are dichotomous—pass/ fail.
- The trials are independent—one trial is not dependent on others.
- The probability of success $p = 0.85$ remains the same for all trials.

Solution:

A trial is taking a polygraph exam. Success = Pass. Failure = Fail. $n = 9$; $p = 0.85$; $q = 0.15.4$

Let X be a random variable denoting the number of students who pass the polygraph examination.

a. $P(x = 9) = 0.232$

b. $$\begin{aligned} P(x \geq 5) &= P(r = 5) + P(r = 6) + P(r = 7) + P(x = 8) + P(x = 9) \\ &= 0.028 + 0.107 + 0.260 + 0.368 + 0.232 = \mathbf{0.995} \end{aligned}$$

c. $P(x \leq 4) = 1 - P(x \geq 5) \approx 1 - 0.995 = \mathbf{0.005}$

d. $P(x = 0) = 0.000$ (to 3 digits)

Case Study 2: Fishing: Northern Pike

Manitoba Northern Pike are hardy, tough fish! using artificial lures with barbed treble hooks, it was found that the hooking mortality rate was only about 5%. This means that only 5% of pike that were caught and released died (Adapted from proceedings of National Symposium on Catch and Release Fishing, sponsored by Humboldt State University). Suppose that a group of anglers caught and released 15 northern pike Manitoba. What is the probability that,

a. None of the fish died?
b. Less than 3 of the fish died?
c. All the fish lived?
d. More than 13 fish lived?

Suitable Tool: Binomial distribution.

Reason:

- The number of trials $n = 15$ is finite.
- The trials are dichotomous—dies/survives.
- The trials are independent—one trial is not dependent on others.
- The probability of success $p = 0.05$ remains the same for all trials.

Solution:

A trial is catching and releasing a pike. Success = pike dies. Failure = pike lives. $n = 15; p = 0.05; q = 0.95$.

Let X be a random variable denoting the number of fish that died.

a. $P\ (x = 0) = \mathbf{0.463}$
b. $P\ (_{x>}3) = 1 - P(X \leq 3) = \mathbf{0.964}$
c. $P\ (x = 0) = \mathbf{0.463}$ (all lives are equal to none die).
d. Change success to live. **p = 0.95, P (x > 13) = 0.828**

Case Study 3: Health Care: Diabetes

People with diabetes may develop other health complications associated with the disease. The following information is based on a feature in USA and today entitled "A Look at Statistics That Shape Our Lives". About 40% of all people with diabetes will also develop hypertension (blood pressure problems), and about 30% of people with diabetes will develop eye disease. Suppose that you are the director of a healthcare centre that has 12 people with diabetes and no other related health problems. Part of your job is to monitor these patients for symptoms of new illness related to diabetes so that corrective measures can be started. What is the probability that,

a. None of the diabetes patients will ever develop related hypertension?
b. Fewer than 4 of the diabetes patients will ever develop related hypertension?
c. No more than 3 diabetes patients will ever develop a related eye disease?
d. At least 5 diabetes patients will never develop a related eye disease?

Suitable Tool: Binomial distribution.

Reason: The researcher can now appreciate why binomial distribution is used.

Solution:

Let X be a random variable denoting the number of diabetes patients.

a. $n = 12, p = 0.40.\ P\ (x = 0) = \mathbf{0.0022}$
b. $n = 12, p = 0.40,\ P\ (_{x<}4) = \mathbf{0.225}$
c. $n = 12, p = 0.30,\ P\ (_{x\leq}\ 3) = \mathbf{0.4925}$
d. $n = 12, p = 0.70;\ P\ (_{x\geq}\ 5) = \mathbf{0.9905}$

How to compute μ and σ for a binomial distribution

μ **= np** is the expected number of successes for the random variable r. $\sigma = \sqrt{\mathbf{npq}}$ is the standard deviation for the random variable r, in which r is a random variable representing the number of successes in a binomial distribution. ***n*** is the number of trials. ***p*** is the probability of success on a single trial, and $q - 1 = p$ is the probability of failure on a single trial.

Case Study: Hype—Improved products

The Wall Street Journal reported that approximately 25% of the people who are told a product is improved will believe that it is in fact improved. The remaining 75%

believe that this is just hype (the same old thing with no real improvement). Suppose a marketing study consists of a random sample of 8 people who are given a sales talk about a new, improved product.

Solution:

$$\mu = n^*p = 8^*0.25 = \mathbf{2}$$

$$\sigma = \sqrt{npq} = \sqrt{8 * 0.25 * 0.75} = \mathbf{1.225}$$

Case study: Criminal justice: Jury duty

Have you ever tried to get out of jury duty? About 25% of those called will find an excuse (work, poor health, travel out of town, etc.) to avoid jury duty (Adapted from Bernice Kanner, Are You Normal? St. Martin's Press, New York). If 12 people are called for jury duty (Fig. 3.8)

•

a. What is the probability that all 12 will be available to serve on the jury?
b. What is the probability that 6 or more will not be available to serve on the jury?
c. Find the expected number of those available to serve on the jury. What is the standard deviation?

Suitable Tool: Binomial distribution.

Reason: The number of trials $n = 12$ is finite.
The trials are dichotomous-serve/do not serve.
The trials are independent-one trial is not dependent on others.
The probability of success $p = 0.75$ remains the same for all trials.

Fig. 3.8 Siméon-Denis Poisson

Solution:

Let X be a random variable denoting the number of people called. $n = 12$; $q = 0.25$ do not serve; $p = 0.75$ serve;

a. $P(x = 12, \text{serve}) = 0.032$
b. $P(x \geq 6 \text{ do not serve}) = 0.053$
c. For serving, $\mu = 9, \sigma = 1.50$

The Poisson distribution

An Architect of the Poisson distribution is **Simeon-Denis Poisson**, A French mathematician who introduced this distribution through his theory which was published in 1837. He was also known for his extensive contributions to definite integrals, electromagnetic theory, and probability.

Poisson Event [7]**:** The Poisson distribution arises from situations in which there is a large number of opportunities for the event under scrutiny to occur but a small chance that it will occur on any one trial.

The Poisson distribution is used when we need to find the probability of how many random events can occur over a period of time. This vital information is captured by Lambda λ. Which is the parameter of the distribution.

Meaning of parameter Lambda: λ is the rate parameter [8]

The Poisson distribution probability mass function gives the probability of observing ***k*** events in a time period given the length of the period and the average events per time.

$P(k \text{ events in time period}) = e^{\left[-\frac{\text{events}}{\text{time}} * \text{time period}\right]} * \left[\frac{\left(\frac{\text{events}}{\text{time}} * \text{time period}^{k}\right)}{k!}\right]$

To simplify this formula, Lambda takes the value of $\frac{\text{Events}}{\text{Time}}$ and so the equation is now explained as follows:

Definition

A random variable X is said to follow a Poisson distribution if it assumes only non-negative values and its probability mass function is given by:

$$P(x, \lambda) = P[X = x] = \begin{cases} \frac{e^{-\lambda} * \lambda^{x}}{x!} & \text{for all values of } x = 0, 1, 2 \ldots \text{ and } \lambda > 0 \\ 0, & \text{Otherwise} \end{cases}$$

Characteristics:

- n is very large.
- p is very small.
- $\lambda = np$ is finite and constant.

Poisson distribution may be successfully employed in the following instances:

- The number of deaths from a disease (not due to an epidemic) such as heart attack or cancer or due to snake bite.

- The number of suicides reported in a particular city.
- The number of faulty blades in a pack of 100.
- The number of air accidents in some unit of time.
- The number of printing mistakes on each page of a book.

Mean: $\mu = E(X) = \lambda$
Variance: $\sigma^2 = V(X) = \lambda$
Standard deviation: $\sigma = \sqrt{\lambda}$
In data analytics, we often employ variables that are in countable units such as the number of teachers leaving the profession in a given year and attendance of students per semester. These are generally measured as averages or treated as interval/ratio variables. This is because distances between various points on these variables are constant, and also, they have a true zero point.

Why is Poisson distribution suitable for count data?

This distribution is skewed, non-negative and suitable for low-frequency count variables, wherein many values are close to zero/initial values (Fig. 3.9).

A famous example of Poisson distribution is to be found in a study: Updating a Classic: **"The Poisson Distribution and the Supreme Court"**. Wallis (1936) studied vacancies in the US Supreme Court over 96 years (1837 -1932) and found that the distribution of the number of vacancies per year could be characterized by a Poisson model. These vacancies have occurred because of the death or retirement of Supreme Court justices. Historically, the proportions of vacancies due to death and vacancies due to retirement have been roughly equal in number. Studying vacancies from 1837 to 1932, Wallis found that a Poisson process with parameter $\mu = \mathbf{0.5}$ vacancies/year gave a remarkably good fit to the distribution of the number of vacancies per year over the sample period.

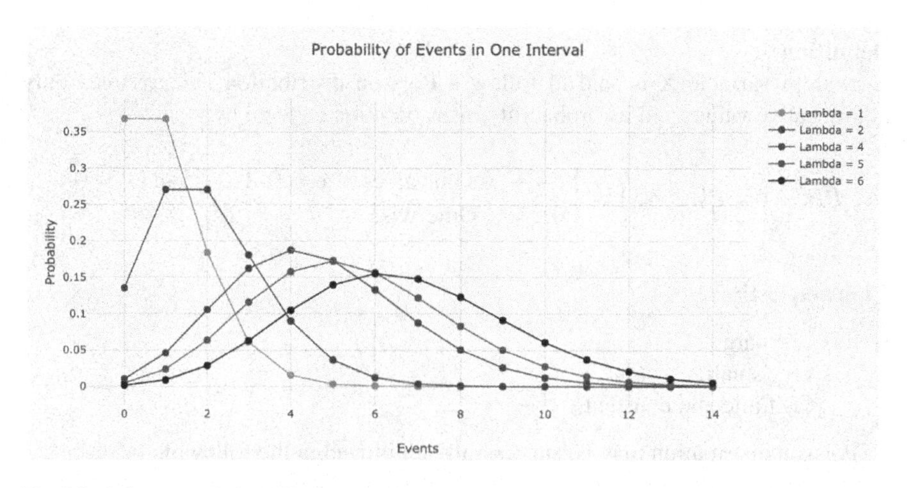

Fig. 3.9 Poisson graph for different lambda values

Poisson was applied originally to the occurrence of criminal cases coming before the courts and the decisions made by judges. This 1838 paper was entitled, "**Research on the Probability of Judgements in Criminal and Civil Matters**".

Question: Geiger–Muller detector uses Poisson distribution. Why?

A Geiger–Muller (GM) detector is an instrument used to detect the ionizing effect (alpha, beta and gamma) of radiation. The radioactive element when hits the central electrode, and a clicking sound is heard, and thus every sound is recorded. The time when these radioactive elements are detected is completely random and unpredictable. An activity is defined as the number of radioactive atoms which disintegrate and emit radioactivity per second.

Activity $= \frac{\text{Numberofdisintegrations(N)}}{\text{Timetaken(t0}} = \frac{\text{N}}{\text{t}}$

Thus, a Poisson process is used to make sense of this randomness. When random events are happening over a defined period of time, $t_0 - t_1$ we define a Poisson process with Lambda as the "**rate of arrivals**".

Example 13 The particles arriving at the detector are defined as a Poisson process with Lambda 0.7/second. What is the probability that the Geiger counter detects 3 or more particles in the next 4 s?

Solution

The number of radioactive particles arriving at the detector in the next 4 s is defined as $\mu = \lambda^* (t_1 - t_0)$ where the time intervals t_1 and t_0 are non-overlapping, independent, short intervals of time.

$\mu = 0.7^* (4–0) = 2.8$

$= P(X \geq 3) = 1 - P(X < 3)$

$= 1 - \{[f(0) - f(1) - f(2)]\}$

$= 1 - [\left(e^{-2.8}\frac{2.8^0}{0!}\right) + \left(e^{-2.8}\frac{2.8^1}{1!}\right) + \left(e^{-2.8}\frac{2.8^2}{2!}\right)]$

$= 1 - [0.0681 + 0.170268 + 0.238375] = \mathbf{0.530546}$

Why Poisson?

When the entire time is split into short, non-overlapping and independent intervals of time, we can also expect the arrivals to follow a binomial distribution with a small probability of ***p*** and the number of segments as ***n***. But this is not right because in such a case each tiny segment of time should either contain 1 or 0 arrivals only. Usually, if Δt is so small that it is rare to get even 1 arrival in a segment, then it is essentially impossible to get 2 arrivals. Therefore, Poisson is the best practice when the arrivals are random and unique.

When to use Binomial and Poisson distribution?

- Binomial distribution is to be used when an exact probability of an event is given or is implied in the situation. It will calculate the probability of that event "***k***" times out of "**n**".
- If a **mean or average** probability of an event happening per unit is given and to calculate a probability of *n* events happening in a given time, the Poisson distribution is used.

Case Study I: A typist makes on average 3 mistakes per page. What is the probability of a particular page having no errors on it?

Suitable Tool: Poisson distribution

Reason: In this situation, the exact probability is not given but an average value $\lambda =$ 3 errors per page. Therefore, Poisson distribution has to be used.

Ans: Let the number of mistakes made by the typist be denoted by the random variable X. Then, $p(x) = \mathrm{e}^{-\lambda}\lambda^x/x!$

$P(X = 0) = p(0) = \mathrm{e}^{-3}3^0/0! = 0.0498$

Example 14 40 components are packed in a box. The probability of a component being defective is 0.1. What is the probability of the box containing 2 defective components?

Suitable Tool: Binomial distribution

Reason: In this case, a definite value of probability is given, namely $p = 0.1$ and thus,

$q = 1 - p = 0.9$. Thus, the binomial distribution is to be used with $n = 40$.

Ans: Let the number of defective components be denoted by the random variable X. Then,

$P(X = 2) = p(2) = {}^5C_2\ (0.1)^2\ (0.9)^{38} = \mathbf{0.1423}$.

Example 15 A computer crashes once every 4 days on average. What is the probability of there being 2 crashes in one week?

Suitable Tool: Poisson distribution

Reason: The average value is known and not the exact number. Due to the uncertainty, Poisson is the best distribution to be used.

Solution

Let the number of crashes in a week be denoted by the random variable X. Given, $\lambda= 4$. Then, $p(x) = \mathrm{e}^{-\lambda}\lambda^x/x!$

$P(X = 2) = p(2) = \mathrm{e}^{-4}4^2/2! = \mathbf{0.1465}$

- A researcher may be interested in knowing if Party A will come into power. Or if a particular drug causes high BP. In such situations where variables are dichotomous, we use binomial distribution.
- One example of a use case of binomial distribution in university research can be a study in which the proportion of girls among all students taking up an advanced computing course is to be studied.
- Let the proportion of girls in a coding institute be equal to 0.467. There are 20 students who opt for a certificate course in advanced computing as a certificate programme. If a researcher wants to estimate the probability of the number of girls in this selection, say the probability that all are girls, two are girls or all are boys, the binomial distribution is the suitable statistical tool.

Example 16 A researcher wants to test children born in a big family. It is already known that the numbers of female children born are twice as many as the number of males in the family till now. He/ she wishes to make a statement on how difficult it is to follow an unwritten rule that key positions in family businesses can be occupied by male children. He intends to show this by finding the probability of 5 male children being born and showing that it is very less. Assuming that there are 5 key positions in the family business, binomial distribution can be used as follows (Fig. 3.10):

Method: Let X be the random variable denoting the number of males in the family.
Given $p = 1/3$, $n = 5$, $q = 2/3$
P (only males) $= P(X = 5)$
$= n_x^c p^x q^{(n-x)}$
$= \mathrm{n}_5^5 \left(\frac{1}{3}\right)^5 \left(\frac{2}{3}\right)^{5-5} = \mathbf{0.00412}$

The very famous Normal Distribution

The normal or the Gaussian distribution is also called the **"Backbone of Data Science"**. As the name suggests, this distribution is the most commonly used continuous frequency distribution which accurately describes the distribution of values.

History of Normal Distribution

The invention of this comprehensive and powerful distribution was credited to French scientist **Abraham de Moivre.**

The normal curve formula first appeared in a paper by DeMoivre in 1733. In the second edition of **"The Doctrine of Chances"**, the expression "***Normal Curve***" was not used till the 1870s. It all started when Moivre was once solving a gambling problem where there were entities that were in binomial distribution format that had to be summed up. Later in 1800, Gauss used the normal distribution formula to describe a pattern of measurement errors in observational data.

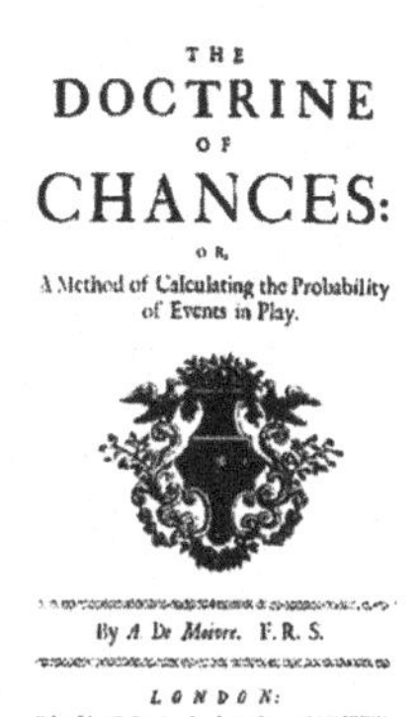

THE
DOCTRINE
OF
CHANCES:
OR,
A Method of Calculating the Probability
of Events in Play.

By *A. De Moivre.* F. R. S.

LONDON:
Printed by *W. Pearson*, for the Author. MDCCXVIII.

Fig. 3.10 Carl Friedrich Gauss

The Normal or Gaussian Distribution

The normal or Gaussian distribution is found by **Johann Friedrich Carl Gauss.** This distribution was studied by the French mathematician Abraham de Moivre (1667–1754) and later by Carl Friedrich Gauss (1777–1855). He was regarded as one of the greatest German mathematicians of all time for his priceless contributions to number theory, probability theory, geodesy and potential theory including electromagnetism. The work of these mathematicians provided a foundation upon which the theory of statistical inference is based upon (Fig. 3.11).

Introduction to Normal Distribution

Let's start with the Unknown X again! (The one we saw at the beginning of the first chapter). If we need to predict X, then we need to know some background information about X. So that our prediction is not baseless. In statistical terms, we term this as "*Understanding the underlying distribution*".

When we understand the distribution of data, we get a general perception of how the data is modelled so that it's easy to estimate the unknown X. One of the most famous distributions in statistics is the **Gaussian or normal distribution.** This is because the majority of data sets available can be modelled into a normal distribution. This distribution is like a symmetric bell-shaped curve with asymptotic tails at the ends.

Fig. 3.11 Abraham de Moivre

A bell curve is like ***a** "One size fits all model "!* This means that when **n (the sample size)** increases most of the data sets tend to fit into a bell curve, this is the idea behind a famous theorem called **"central limit theorem"** defined as,

> The central limit theorem (CLT) states that the distribution of a sample variable approximates a normal distribution (i.e. a 'bell curve') as the sample size becomes larger, assuming that all samples are identical in size, and regardless of the population's actual distribution shape [12].

To get a better understanding of C.L.T, take a look at Fig. 3.12 the first row shows the shape of a few data models like the uniform distribution, skewed normal curve and a general random pattern of data. In the second row, the shape of these data sets is depicted when the number of data points is just two units. In the third row, we can notice that as the sample size increases ($n = 30$) all the data types tend to take a shape of a bell curve. This is the concept behind the central limit theorem. This is a vital theorem having its applications in all fields of analytics.

Graph of a Normal Distribution:

Let's recollect and collate all that we learnt about normal distribution.

1. The curve is perfectly symmetrical about the mean of the distribution. As a result, the mean of the distribution is the same as the mode and the median.
2. The curve approaches the horizontal axis but never touches or crosses it.
3. The transition points between cupping upward and downward occur above and below these interval ranges, $\mu + \sigma$ and $\mu - \sigma$.

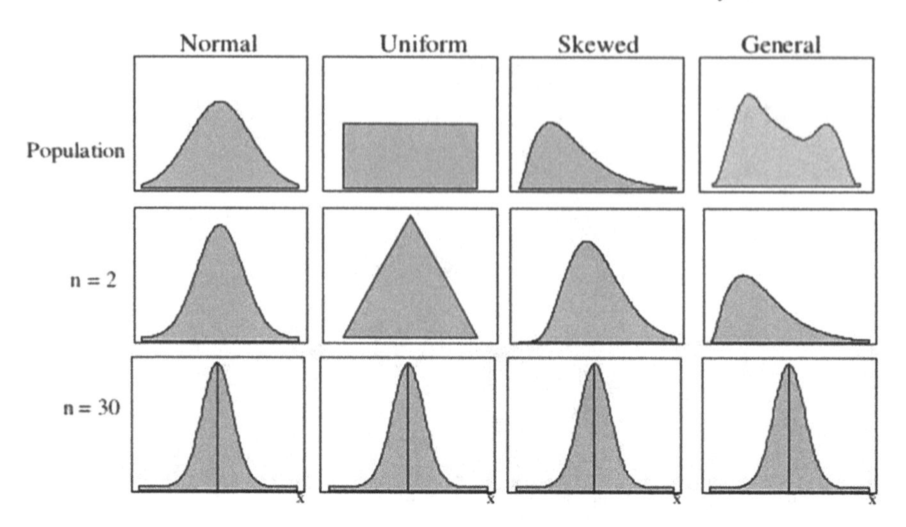

Fig. 3.12 Diagrammatic explanation of (C.L.T)

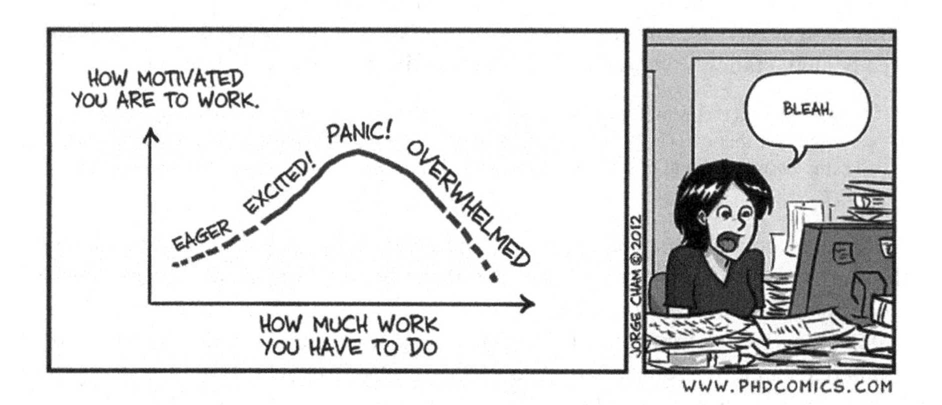

A note on Moments, Skewness and Kurtosis of Data

When unknown data is modelled, statisticians use a tool called "Moments" to understand different aspects of this distribution of data. There are generally 4 moments that are popularly used.

Definition
Moments are a set of statistical parameters to measure a distribution.

Moment 1: This defines the mean of the distribution μ_1

Moment 2: The variance of data μ_2

Moment 3: Talks about the skewness of the data μ_3

Moment 4: Defines the kurtosis of the data μ_4

Moments are broadly classified into raw or uncentered moments and central moments. These differ in the type of calculations. This can clearly be understood by Fig. 2.20 more about moments and moment generating functions will be discussed in the chapters ahead.

Generally, in statistics, the third-order central moment is defined as skewness.

Skewness and Kurtosis

If you ever had to define how a distribution looked, you would definitely talk about its shape and about the height of the curve. That's exactly what skewness and kurtosis refer to. A normal distribution is perfectly symmetric with skewness 0 and kurtosis of 3.

Skewness and its types: Skewness defines the symmetry of data.

Left Skew—Negative skew:

- This distribution is skewed to the left.
- The mean is pulled to the right from the centre.
- Data that is skewed to the left has a long tail that extends to the left; that is, it is negatively skewed (Fig. 3.13).
- In this situation, the mean and the median are both less than the mode.
- In general, for data skewed to the left, the mean will be less than the median.

Right Skew—Positive skew:

- This distribution is skewed to the right.

Fig. 3.13 Raw and central moments

Moment	Uncentered	Centered
First Order	$E(X) = \mu$	
Second Order	$E(X^2)$	$E((X - \mu)^2)$
Third Order	$E(X^3)$	$E((X - \mu)^3)$
Fourth Order	$E(X^4)$	$E((X - \mu)^4)$

Mean	$E(X)$
Variance(X)	$E((X - \mu)^2) = \sigma^2$
Skewness (X)	$E((X - \mu)^3) / \sigma^3$
Kurtosis(X)	$E((X - \mu)^4) / \sigma^4$

- The mean is pulled to the left from the centre.
- Data that is skewed to the right has a long tail that extends to the right; that is, it is positively skewed.
- In this situation, the mean and the median are both more than the mode.
- In general, for data skewed to the right, the mean will be less than the median.

Symmetrical Distribution—Perfect Skew: A perfectly symmetric data has

- Zero skewness.
- Mean = Median = Mode.
- Quartiles are equidistant from the median.

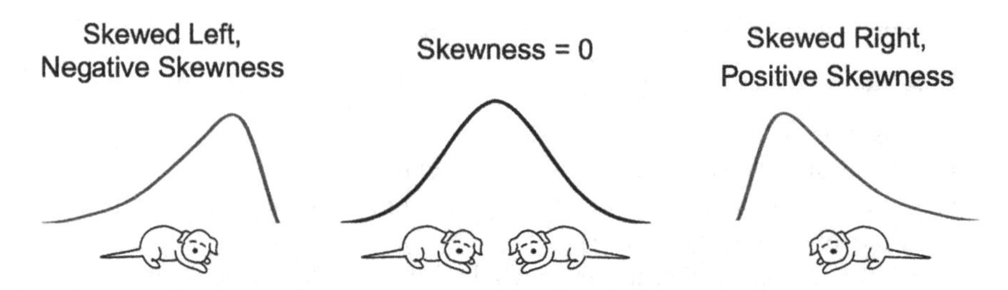

Reproduced by permission of John Wiley and Sons. From the book, Statistics from A-Z—Confusing Concepts Clarified.

The coefficient of skewness is a measure that indicates the degree of skewness. It may be positive, zero or negative. It will be a positive value, if the right tail of the distribution is longer than the left tail, a negative value, if the left tail is longer than the right tail and zero if the distribution is symmetrical.

Karl Pearson's coefficient of skewness is given by:

$$S = \frac{\text{Mean - Mode}}{\text{Standard Deviation}} = \frac{\overline{x} - Z}{\sigma}$$

When the mode is ill-defined, Karl Pearson's coefficient is used and is defined as

$$S = \frac{3(\text{Mean - Median })}{\text{Standard Deviation}} = \frac{3(\overline{x} - M)}{\sigma}$$

Bowley's coefficient of skewness which is based on quartiles is given by:

$$Sk_B = \frac{Q_3 + Q_1 - 2Q_2}{Q_3 - Q_1}$$

Kurtosis

Kurtosis gives us an idea about the "flatness or peakedness" of the frequency curve. Kurtosis defines the **"Convexity of the Frequency Curve"**.

Sample Kurtosis Formula:

$$K = \frac{n(n+1)(n-1)}{(n-2)(n-3)} \frac{\sum_{i=1}^{n} \left(X_i - X_{\text{avg}}\right)^4}{\left(\sum_{i=1}^{n} \left(X_i - X_{\text{avg}}\right)^2\right)^2}$$

Population Kurtosis Formula:

$$K = n \frac{\sum_{i=1}^{n} \left(X_i - X_{\text{avg}}\right)^4}{\left(\sum_{i=1}^{n} \left(X_i - X_{\text{avg}}\right)^2\right)^2}$$

It is measured by the coefficient β_2 which is given by:

$\beta_2 = \frac{\mu_4}{\mu_2^2}$, where $\mu_2^{'}$ is the second moment about the origin and μ_4 is the fourth-order central moment.

- If $\beta_2 = 3$, the distribution (curve) is **Mesokurtic or normal.**
- If $\beta_2 > 3$, the curve is more peaked than the normal curve and is said to be **Leptokurtic**.
- If $\beta_2 < 3$, the curve is flatter than the normal curve and is said to be **Platykurtic**.

Kurtosis for normal distribution is 3. This means the bell curve is **Mesokurtic**.

The famous 3 Sigma Rule of Normal Distribution[1]

In Chap. 2, we saw a dinosaur example for one standard limit from the mean. Let us understand what is meant by, 1, 2 and 3 standard deviations from the mean in normal distribution in detail (Fig. 3.14).

Standard deviations will be the area enclosed between limits:

Upper Limit: Mean + (2 * Standard deviation).

Lower Limit: Mean − (2 * Standard deviation).

Similarly,

Standard Deviations Will Be the Area Enclosed Between Limits:

Upper Limit: Mean + (3 * Standard deviation).

Lower Limit: Mean − (3 * Standard deviation).

Now that we have understood the concept of standard deviations from the mean, we can understand this popular rule in a normal curve in an easier manner. Have a look at Fig.—, we can observe that in a normal curve,

[1] https://stats.stackexchange.com/questions/476677/understanding-standard-deviation-in-normal-distribution

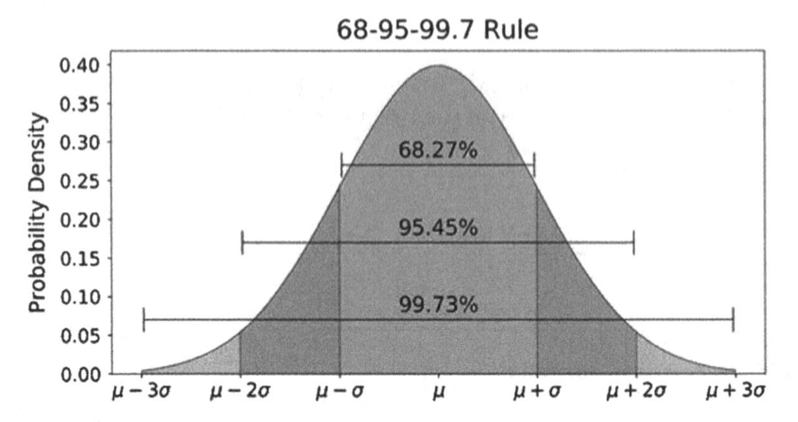

Fig. 3.14 Standard deviations under 1, 2 and 3 sigma limits

- **68.3%** of data is included under the one standard deviation limit (*More than 50% of data*).
- **95%** of data is included under two standard deviation limits.
- **99.7%** of data is included under the 3 standard deviation limits.

This is another reason why the normal curve is very popular, it includes almost 100% of all data points within its 3 standard deviation limits.

The Normality Check: Is your data Normal?

It is always a prerequisite to check for normality of data as this is an underlying assumption for all parametric testing, ANOVA, regression and correlation too. With large sample sizes (>30 or 40) the normality assumption does not make a significant impact because of the central limit theorem (CLT). Yet it is very important to ascertain whether the data shows a serious deviation from normality before we conclude any statistical results.

Visual Tools:

Q-Q plot (quantile–quantile plot), histograms, box plots and stem and leaf plots check for normality through symmetry, bell shape and outliers in data. These tools are only for immediate visual judgement. But this approach is unreliable and not a suitable practice for all data sets.

To overcome this disadvantage, statistical tests such as Kolmogorov–Smirnov (K-S tests) and Shapiro–Wilk tests are used for reliable checks. In both the test cases, if the *p*-value is greater than 0.05, this indicates that the data is normally distributed. (Testing procedures are discussed in the next chapter in detail.)

Why are these concepts important?

1. The two key parameters mean μ and standard deviation σ can define the shape and location of the normal distribution. The symmetry can be obtained only when the three central tendencies (mean, mode and median) are equal.

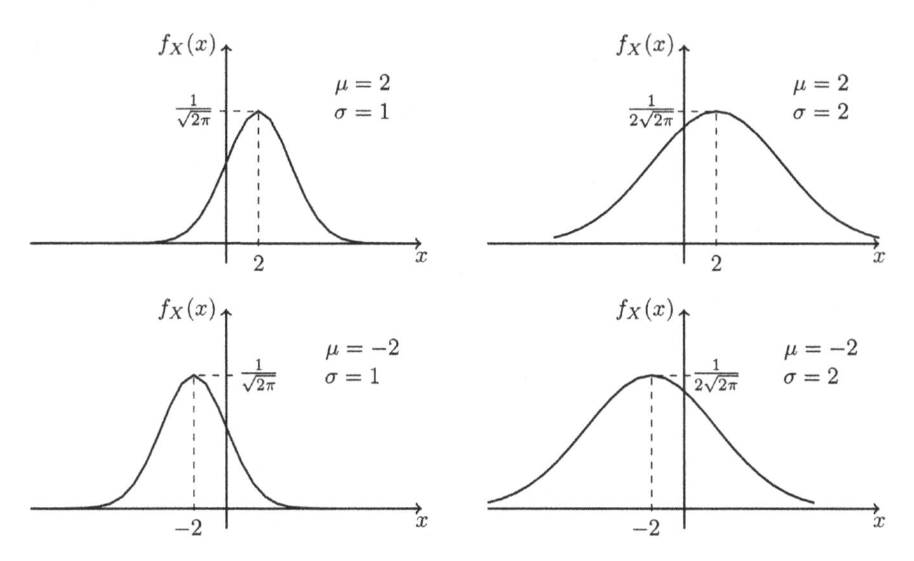

Fig. 3.15 Change in the mean and standard deviation of normal distribution [9]

- Change in the mean- Implies shifting the curve along the number line, while
- Change in the standard deviation impacts the stretch of the curve.

2. Using the empirical rule, we include close to 98% of data in the bell curve, which means the rest of the points can be considered as signal noise or outliers which don't belong to the distribution.

Definition

A random variable *X* is said to have a normal distribution with parameters μ (mean) and σ^2 (variance) if its probability density function (pdf) is given by the probability law (Fig. 3.15).

$$\mathbf{f(x;\, \mu, \sigma)} = \frac{\mathbf{1}}{\boldsymbol{\sigma\sqrt{2\Pi}}}\mathbf{exp}\left\{-\frac{\mathbf{1}}{\mathbf{2}}\left(\frac{\mathbf{x}-\boldsymbol{\mu}}{\boldsymbol{\sigma}}\right)^2\right\}, -\infty < \mathbf{x} < \infty, -\infty < \boldsymbol{\mu} < \infty, \boldsymbol{\sigma} > \mathbf{0}$$

Case Study 1: Medical—Blood Glucose

A person's level of blood glucose and diabetes are closely related. Let *x* be a random variable measured in milligrams of glucose per decilitre (1/10 of a litre) of blood. After a 12-h fast the random variable *x* will have a distribution that is approximately normal with mean $\mu = 85$ and standard deviation $\sigma = 25$ (diagnostic test with nursing implications, edited by S. Noeb, Springhouse Press).

Note: After 50 years of age both the mean and standard deviation tend to increase. What is the probability that, for an individual (under 50 years) the values increase after a 12-h fast?

a. x is more than 60?
b. x is greater than 60?
c. x is between 60 and 110?
d. x is greater than 140 (borderline diabetes starts at 140)?

Suitable Tool: Normal distribution

Solution:

Using the standard normal distribution (z) table, the probabilities are obtained as follows:

a. $P(X > 60) = P\left(\frac{x-\mu}{\sigma} > \frac{60-85}{25}\right) = P(Z > -1) = \mathbf{0.8413}$
b. $P(X < 110) = P\left(\frac{x-\mu}{\sigma} < \frac{110-85}{25}\right) = P(Z < 1) = \mathbf{0.8413}$
c. $$\mathrm{P}(60 \leq X < 110) = P\left(\frac{60-85}{25} \leq \frac{x-\mu}{\sigma} \leq \frac{110-85}{25}\right)$$
$$= P(-1 \leq Z \leq 1) = 0.8413 - 0.1587 = \mathbf{0.6826}$$
d. $P(X > 140) = P\left(\frac{x-\mu}{\sigma} > \frac{140-85}{25}\right) = P(Z > -2.20) = \mathbf{0.0139}$

The Standard Normal Distribution

This is a normal distribution written as $N(0, 1)$ which means, the parameters (mean and standard deviation) are by default given values 0 and 1, respectively (Fig. 3.16). $\mu = 0$ and $\sigma = 1$.

Z-scores and T-scores:

Consider an example of two students A and B in different sections. A got a score of 70 whereas the mean score of his/her section was 60. B got 80 whereas his section's mean score was 70. This information tells that both A and B were 10 marks above the average of their respective sections but does not speak about the spread of the distribution, in other words, how the person scored with respect to others in the class.

The scores are converted into normalized or z-scores so that a standard scale is obtained to help in comparison. The process of converting or transforming scores of a variable to Z-scores is called data standardization.

Note:

- **Z-Scores** are a transformation of raw scores into a standard form, where the transformation is based on knowledge about the population's mean and standard deviation.
- **T-Scores** are a transformation of raw scores into a standard form, where the transformation is made when there is no knowledge of the population's mean and standard deviations are computed by using the sample's mean and standard deviation, which is the best estimate of the population's mean and standard deviation.

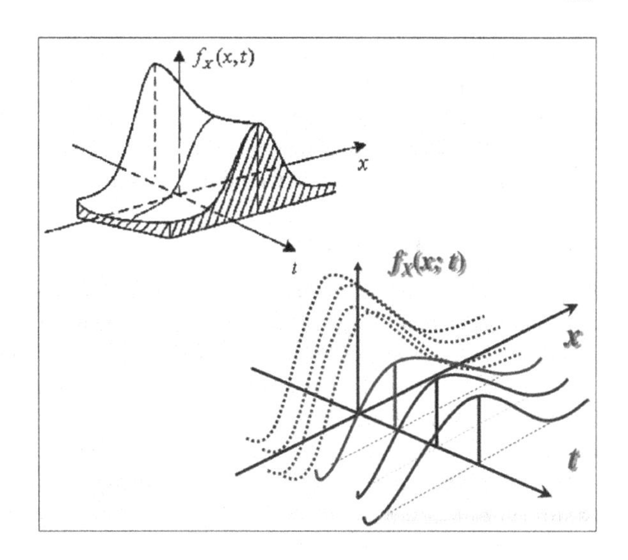

Fig. 3.16 Independent and identically distributed variables

Independent and Identically Distributed Variables (I.I.D) [10]

This is an important assumption of the central limit theorem. Independent and identical distribution means that in the random process, if these random variables obey the same distribution and are independent of each other, then these random variables are called IID variables. Figure 3.16 gives a diagrammatic representation of the same. If the random variables X_1 and X_2 are IID variables, then the value of X_2 does not affect the value of X_1 and both the random variables X_1 and X_2 obey the same distribution, for example, if the experimental conditions remain unchanged, then the positive and negative results of a series of coin tosses are independent and identically distributed.

Data standardization

Converting raw scores into standard normalized scores (Z- scores):

When data is available on different kinds of assessment tests, we cannot compare those scores directly, unless and of course, the scales of these two tests are the same.

We can overcome the problem by standardizing the scores. We convert the raw scores to standard normalized scores by using the relation $z = \frac{x-\mu}{\sigma}$.

This clearly indicates how many standard deviation units a raw score is above or below the mean, thus providing a standard scale for comparison.

Note:

1. Converting ***x***-scores to z-scores will not change the shape of the distribution. However, the distribution of z-scores is normal if and only if the distribution of x is normal.
2. If **z** values can be negative or in decimals, they are converted into t values by multiplying with some constants and added to some other constants. **For example, t = 10z + 40.**

3. If we sum up all the z values, we will get 0 as the positive and negative scores will nullify each other.
4. These z-scores are equally distributed around the mean of the distribution.
5. When the distribution is skewed, more positive z-scores exist in right-skewed data and similarly, left-skewed data will have more negative z-scores.

Comparisons using Z-scores

An important aspect of data analysis is when data sets need to be compared on different parameters to arrive at a reliable conclusion. Standardizing the data sets makes it easier for comparison as the data sets are brought to equal levels.

For example, consider the scores of 2 brothers who wrote an entrance examination at different years. Elder brother scored 1850 with an overall mean and standard deviation of 1500 and 300, respectively. As time passed, the evaluation method changed and the younger brother scored 25 when the performance of students had a mean and standard deviation of 21 and 5, respectively. In this case, standardization makes the values stand on the same ground which can thus be compared.

Z (Elder brother) = (1850–1500)/300 = **1.16 standard deviation above the mean.**

Z (Younger brother) = (25–21)/5 = **0.8 standard deviation above the mean.**

This way we can compare and conclude that the elder brother scored better than the younger brother.

Case Study: Honolulu Temperature: Data collected over years show that the average temperature of Honolulu is $\mu = 73$ °F standard deviation $\sigma = 5$ °F (U.S. Department of Commerce, Environmental Data Service). Convert each of the following °F intervals into an interval of z values.

Example 17 Convert the following intervals of z values to intervals in degree Fahrenheit.

1. $53\ °F < x < 93\ °F$
2. $x < 65\ °F$
3. $78\ °F < x$
4. $1.75 < z$
5. $z < -1.90$
6. $-1.80 < x < 1.65$

Suitable Tool: Standard Normal Distribution

Solution

1. $53^{O}F < x < 93^{O}F = P(\frac{53-73}{5} < \frac{X-\mu}{\sigma} < \frac{93-73}{5}) = -4 < z < 4$
2. $X < 65°F = (\frac{x-\mu}{\sigma} < \frac{65-73}{5}) = z < -1.6$
3. $78°F < x = (\frac{78-73}{5} < \frac{x-\mu}{\sigma}) = 1 < z$
4. $1.75(5) + 73 < z\sigma + \mu = 81.75°F < x$

5. $z\sigma + \mu < -1.95(5) + 73 = x < 63.5°F$
6. $-1.8(5) + 73 < z\sigma + \mu < 1.65(5) + 73 = 64°F < x < 81.25°F$

The probabilities of all these results can be found using the Z normal table

Case study: A student obtains 85 marks in Maths and 60 in English. If the mean and standard deviation of scores are 65 and 15 for math and 50 and 5 for English, find the subject in which he scored better.

A layman would probably conclude that Math is better as he scored 25 more, on average in that subject. But this will be erroneous as standard deviation is also in the picture. The right way would be to convert these scores into z-scores as follows:

Z-score in math $= \frac{85-65}{15} = \mathbf{1.33}$

Z-score in English $= \frac{60-50}{5} = \mathbf{2}$

Calculating the probabilities of these values. Thus, he has done better in English than in math, as is obvious from the z-scores.

Note: In educational and psychological testing, the normal distribution is used to determine the relative difficulty of test questions. Consider a paper to have 3 subdivisions A, B and C. Let the number of students who answered section A correctly be 25% which means, 75% are not able to correctly answer it. Similarly, let 10% and 20% be the number of people who have answered sections B and C correctly, and thus, 90% and 80% have found sections B and C tough. We assume that the ability measured by these three items is the same, and it is normally distributed.

Scenarios: The score for each student in a class is used to calculate the mean marks which are equal to 60 and SD of 10. What can we say about a student?

(i) With a score of 60?

The z-score is (60–60)/10 = **0**

Interpretation: student score is 0 distance (in units of standard deviations) from the mean, so the student has scored the average marks.

(ii) With a score of 70?

The z-score is (70–60)/10 = **1**

Interpretation: student has scored above average - a distance of 1 standard deviation above the mean.

(iii) With a score of 69.6?

The z-score is (75.6–60)/10 = **1.56**

Interpretation: student has scored above average—a distance of 1.56 standard deviation above the mean.

A note: The normal distribution is vital and omnipresent in all fields of study. But it is essential to recognize that not all human attributes or behavioural events are normally distributed.

Many phenomena display extremely skewed distributions with long upper tails. Examples include the distributions of annual income across households in the country, the output of journal articles by researchers doing Ph.D., the number of violent acts committed by college students, etc.

What is to be done then?

1. Use an appropriate data transformation, for instance, a lognormal distribution becomes normal after a logarithmic transformation. But still, many important variables cannot be normalized in this way.
2. The researchers may use statistics based on various non-parametric or distribution-free methods.

Normal distributions occur when multiple causal processes are additive, whereas the non-normal ones occur when those processes are multiplicative.

Why is normal distribution important?

(i) Because certain theoretical distributions, such as the distribution of possible means, can be very close to normal even when the population distributions are not normal.
(ii) By using the areas underneath normal distributions, we can calculate probabilities of different outcomes, including how likely it is to obtain a mean within a certain range.
(iii) **The central limit theorem (C.L.T)** is an important reason why normal distribution is used in a lot of analytics. Many measurements can be estimated by the normal distribution by assuming that many small, independent effects are additively contributing to each observation.
(iv) The normal distribution has a relationship to least-squares estimation which is one of the simplest methods of statistical estimation
(v) Many psychological, financial and educational variables are distributed approximately normally. Measures of reading ability, introversion, job satisfaction, and memory are among the many psychological variables approximately normally distributed.
(vi) Many statistical tests are derived from the normal distribution and most of them work under the assumption of normality.

Chebyshev's Inequality

Named after a popular Russian Mathematician **Pafnuty Lvovich Chebyshev,** who was well known for his contributions to number theory. He was also very interested in the field of mechanics. He is popularly known for his invention of the very famous orthogonal polynomials.

Definition

If X is a random variable with mean μ and variance σ^2, then for any positive number k, we have

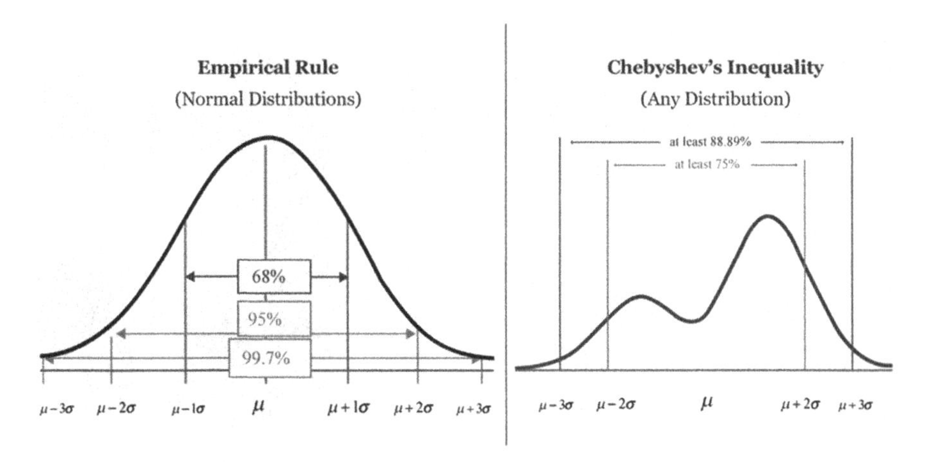

Fig. 3.17 Empirical rule versus Chebyshev's inequality

$$P\{|X-\mu| \geq k\sigma\} \leq \frac{1}{k^2} \text{ or } P\{|X-\mu| < k\sigma\} \geq 1-\frac{1}{k^2}$$

For example, when certain data cannot be modelled to a normal distribution, we need some rule to replace the empirical rule of normal distribution. Chebyshev's Inequality is that rule which gives the minimum percentage of data that must fall within the specified standard deviation of the mean.

Take a look at Fig. 3.17 [13] for a better understanding of the rule. Chebyshev's inequality interprets the role of standard deviation as a parameter to characterize variance. It can be used to express the concept of data spread about the mean for all kinds of distributions, skewed, symmetric or any other shape.

Results of Chebyshev's theorem: For any set of data,

- At least 75% of the data fall in the interval from $\mu - 2\sigma$ to $\mu + 2\sigma$.
- At least 88.9% of the data fall in the interval from $\mu - 3\sigma$ to $\mu + 3\sigma$.
- At least 93.8% of the data fall in the interval from $\mu - 4\sigma$ to $\mu + 4\sigma$.

Example 18 A teacher keeps a tally of how much time students spend in the library. He chooses a random sample of students. Let x be the number of hours a student spends in the library and let $x = 10$ h and standard deviation $= 1.5$ h. Find an interval in which at least 75% of the students will fit (Fig. 3.18).

Solution

According to Chebyshev's inequality, at least 75% of the data must fall within 2 SD of the mean. So the interval is $\mu - 2\sigma$ to $\mu + 2\sigma = 7, 13$.

This means at least 75% of the students spend time between 7 and 13 h in the library.

Fig. 3.18 Pafnuty-Lvovich-Chebyshev

Additional Notes

Data Collection and its methods

Types of Data are classified based on their sources

Primary data: This is original data that has been collected specially for a purpose in mind. It is first-hand data collected directly from the original source- the respondent. The best example of this is the First Information Report (F.I.R) filed in the police station which is a primary source of data for investigation, interviewing customers directly to determine their satisfaction with their Internet Service Provider (I.S.P) is also an example of primary data.

Secondary data: This is data that has already been collected by others and is readily available from other sources. Therefore, secondary data is data that has been collected for purposes other than the problem at hand. Examples available in published newspapers, journals, magazines, unpublished and published research articles, technical reports, published electronic sources, government records, public sector records, etc.

Methods of Data Collection (Primary Data)

- Direct personal interview.
- Indirect personal interview.
- Through respondents.
- Through a mailed questionnaire.
- Through enumerators.
- Surveys.
- SMS.

Advantages of sample studies

The following are the factors that favour sampling:

(i) Lower costs.
(ii) Samples have lower error rates as they can be controlled in a better manner.
(iii) Samples help in decision-making at a faster rate.
(iv) Samples produce more detailed information.
(v) When testing of a sample destroys the units under study, sampling is the only option.

Before you draw a sample

Correct methods must be employed before drawing any conclusions from a sample. It is important to choose the right sample so that it is representative of the population. This will make the analysis effective.

1. Smaller sample sizes have a greater risk of not being representative.
2. Sources of bias should be eliminated.
3. Choose the right sampling techniques according to the study.

Random Sample

Random samples provide an equal chance for each unit of the population to be selected for study, thus they are unbiased. They provide good estimates of the population.

Some research articles use normal distribution:

1. **Title**: Genetic algorithms and Gaussian Bayesian networks to uncover the predictive core set of bibliometric indices.

We use Gaussian Bayesian networks to solve this problem and discover multivariate relationships among bibliometric indices. These networks are learnt by a genetic algorithm that looks for the optimal models that best predict bibliometric data. Results show that the optimal induced Gaussian Bayesian networks corroborate previous relationships between several indices, but also suggest new, previously unreported interactions.

Journal of the Association for Information Science and Technology Volume 67, Issue 7**.** By, Alfonso Ibáñez, Rubén Armañanzas, Concha Bielza, Pedro Larrañaga. First published: 05 May 2015

2. **Journal**: Social Science Research, Volume 41, Issue 1, January 2012, Pages 199–202

Title: Happiness is not normally distributed: A comment to Delhey and Kohler
Author: Wim Kalmijn. Erasmus University Rotterdam, The Netherlands

Delhey and Kohler assume that the happiness distribution at the population level is essentially normal, but this is distorted by the fact that happiness is measured in samples using scales that are discrete and two-sided bounded. This assumption is tested using the probability method and rejected.

3. Journal: Social Science Research, Volume 41, Issue 3, May 2012, Pages 731–734

Title: Happiness inequality: Adding meaning to numbers – A reply to Veenhoven and Kalmijn

Author: Jan Delhey, Jacobs University Bremen, School of Humanities and Social Sciences, Campus Ring 1, D-28759 Bremen, Germany

Ulrich Kohler, Social Science Research Centre Berlin (WZB), Research Unit Inequality and Social Integration, Reichpietschufer 50, D-10715 Berlin, German

4. Beyond mean rating: Probabilistic aggregation of star ratings based on helpfulness

Wenyi Tay, Xiuzhen Zhang, Sarvnaz Karimi

Journal of the Association for Information Science and TechnologyVolume 71, Issue 7 First published: 04 October 2019

Practice Data sets

For R programmers:

1. Learning Distributions Using Titanic Data Set:

https://www.kaggle.com/hamelg/intro-to-R-part-22-probability-distributions

2. Conditional Probability and Bayes Theorem Using IRIS Data Set

https://www.kaggle.com/hamelg/intro-to-R-part-22-probability-distributions

3. Standard Normal Distribution Using "student's" Data Set

https://www.geo.fu-berlin.de/en/v/soga/Basics-of-statistics/Continous-Random-Variables/The-Standard-Normal-Distribution/The-Standard-Normal-Distribution-An-Example/index.html

For Python Programmers

1. Different Distributions and Their Graphs, Concept of Normal Approximation to Binomial, Random Sampling and Cumulative Distribution Functions.

- https://github.com/nakatsuma/probability_and_statistics/blob/master/notebook-b/probability_distribution.ipynb
- https://github.com/nakatsuma/probability_and_statistics/blob/master/notebook-b/probability_distribution.ipynb

References and Research Articles

1. Illustration on impossible events – PHD Comics by Cristi Carlstead.
2. Classical definition of probability: Wikihow: https://www.wikihow.com/Calculate-Probability
3. Permutations and combinations: https://cetking.com/permutation-combination-2/
4. Case study on lie detector test: Adapted from: www.uwlax.edu
5. Stock prices and interest rates: Adapted from Article from http://www.investopedia.com

6. The psychology deceit: Adapted from: www.algebra.com
7. Poisson event: https://www.sciencedirect.com/topics/biochemistry-genetics-and-molecular-biology/poisson-distribution
8. Meaning of Lambda: https://towardsdatascience.com/the-poisson-distribution-and-poisson-process-explained-4e2cb17d459
9. Change in mean and standard deviation: https://towardsdatascience.com/tagged/normal-distribution?p=562b28ec0fe0
10. I.I.D Variables: https://www.codetd.com/en/article/12879978
11. Conditional probability concepts used on Sudden Infant Syndrome (SISD) URL: http://plus.maths.org/content/os/ issue21/features/Clark/index.
12. Central Limit Theorem Definition: https://www.investopedia.com/terms/c/central_limit_theorem.asp
13. Empirical formula Vs Chebyshev's inequality https://calcworkshop.com/joint-probability-distribution/chebyshev-inequality/

Chapter 4
Hypothesis Testing

WHAT

Concepts of hypothesis.
Parametric and Non Parametric Tests.
ANOVA, Post Hoc Tests, Confidence intervals and estimation.

WHY

- To decide if experimental results are due to chance or a specific cause.
- To evaluate the strength of evidence.
- To determine a reasonable estimate when values are unknown.

HOW

- Learning through various examples.
- Understanding the concepts and solving case studies by applying it in real-life scenarios.

WHEN

- To draw inferences about the population using sample.
- To test the value of a population parameter.

S. Prasad, *Elementary Statistical Methods*,
https://doi.org/10.1007/978-981-19-0596-4_4

CHAPTER 4

- Hypothesis testing.
- Parametric testing.
- NonParametric testing
- ANOVA
- PostHoc tests.
- Estimation and confidence intervals.

WITH MR.STAT

WITH MISS TICS

- Case studies and examples for testing procedure.
- Data set on air quality index in Indian cities.
- Predictive modelling for car prices.
- Data set on Northern winds to understand the use case of different testing procedures.
- Hypothesis testing with lens correction data.

4.1 Introduction: Hypothesis

Let's look at the dictionary definition of what a hypothesis is, "A supposition or proposed explanation made based on limited evidence as a starting point for further investigation".

This simply means a hypothesis is a statement of fact/s that has to be tested. Generally, these statements are more specific because they are based on an observed phenomenon or some prior knowledge of the data.

Examples of hypothesis statements

1. More than 60% of students are sleepy in the class post-lunch break.
2. This year, the judiciary system has procrastinated on justice dispensation more than the expected wait time.
3. On a Monday morning, more than half of the staff arrive late to the office when compared to Wednesday.
4. You are a liar! You always have been.
5. More light and less humidity help pea plants to grow better.
6. Earth is round (Fig. 4.1).

A statistical hypothesis: is defined as "A statement or assertion about a population or the probability distribution characterizing a population, which we want to verify, based on information available from a sample".

Fig. 4.1 Reality and co-incidence

Types of hypotheses: There are two types of hypotheses, namely null and alternative hypothesis.

Null hypothesis—$\mathbf{H_0}$.

Also known as **a statement of "no difference"**. Testing the null hypothesis can tell you whether your results are due to the effect of manipulating the dependent variable or just due to a random chance. In this statement, we have the following conditions:

- The variables under study have no relationship with each other.
- The treatments have zero impact.
- The controlling variable does not affect the dependent variable.
- Two groups of data are similar to each other.
- A statement of no bias.
- Has an equality sign.

Alternative hypothesis—$\mathbf{H_1}$

A statement that prefers a new theory to the existing one. **This is the opposite of the null hypothesis** that the researcher sets out to investigate. In this statement, the following conditions are present:

- The variables under study have a defined relationship with like >, <, $\neq$, $\leq$ and $\geq$, etc.
- The treatments have an unknown/known impact.
- The controlling/independent variable affects the dependent variable.
- Two groups of data are not similar to each other.

Take a note of the null and alternative hypotheses that are formulated in different scenarios.

Example 1 We want to test whether the mean GPA of students in American colleges is different from 3.0. The null and alternative hypotheses are:

H_0 The mean GPA of students in American colleges is 3.0. $\mu = 3.0$.
H_1 The mean GPA of students in American colleges is not 3.0. $\mu \neq 3.0$.

Example 2 In the primary elections that were conducted in the city, the election commissioner feels that not more than 30% of the population utilized their right to vote.

H_0 Not more than 30% of the registered voters in the city voted in the primary election $p \leq 30$.
H_1 More than 30% of the registered voters in the city voted in the primary election $p > 30$.

Example 3 Is there a significant difference in marks of students from groups A and B if we give group A the study material and group B the guideline?

H_0 There is no significant difference between the two groups A and B. In other words, any form of guidance given has no effect.

H_1 There is a significant difference between the two groups A and B. In other words, any form of guidance given has an effect on the marks of students.

So, what do we do with the hypothesis?

A **test of statistical hypothesis** is a two-action decision problem after the experimental sample values have been obtained, we either **accept or reject** the hypothesis under certain considerations.

Why is alternative hypothesis important?

The alternative hypothesis holds the idea of the research or analysis, whereas the null hypothesis is a statement that accepts the current situation.

Why is hypothesis so important in research?

1. The hypothesis suggests the type of data that needs to be collected.
2. Hypothesis guides by providing the direction to analyse the data.
3. Hypothesis helps in interpreting and concluding the results.

Basic Terminologies

1. **Test Statistic**: The test statistic is the instrument we use in order to make a decision to reject the null hypothesis. We convert the observed values from the sample into a standard score, assuming that the null hypothesis is true. It is based on an appropriate probability distribution.

2. **Acceptance and Rejection Region** [1]

Let us assume that the data which we are testing follows a normal distribution. Under the null hypothesis, we divide the bell curve into regions that are called **acceptance and rejection regions**. Figure 4.2 depicts the rejection and non-rejection regions. If the test statistic lies in the rejection region, we reject the null hypothesis. Thus, the rejection region is the region of the standard normal curve, corresponding to a predetermined level of significance. The acceptance region is the region under the normal curve that is not covered by the rejection region.

3. **Critical value**

The value that separates the acceptance and rejection region is called **the critical value**.

4. **Level of Significance**

The quantity α is called the level of significance. This value tells us the probability that a result is due to chance. For instance, if the level of significance is at 5%, it means that the researcher is 95% confident that the result is significant, but there is a 5% chance that the result in reality is due to chance. In other words, there is a 5% chance that a result which a researcher has concluded as significant is not significant in reality. Alpha α is also called a type I error (Explained a little later in the same chapter).

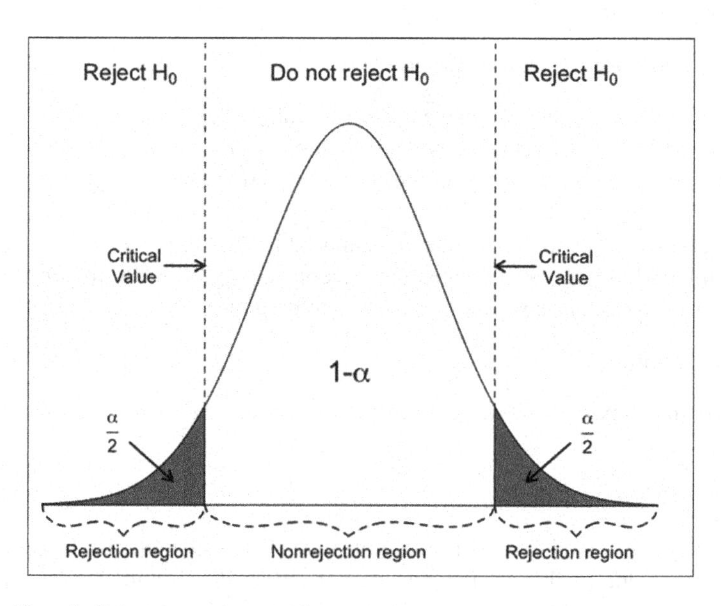

Fig. 4.2 The rejection and acceptance region

5. One-tailed and two-tailed tests

One tail: As the name suggests, in such tests, the rejection region is seen only to one side of the normal distribution curve. In this type of testing, we can check if our statistic is greater than OR less than parametric values **but not both**.

Thus, throughout these tests, we can determine the effect of change in the variable only in a single direction. Therefore, there are again two types of one-tailed tests, which are

- **Left-tailed**: This testing type is used when the alternative hypothesis is defined in terms of the ***Lesser than*** parameter type, which means the null hypothesis is rejected if the test statistic is too small.
- **Right-tailed**: This testing type is used when the alternative hypothesis is defined in terms of the ***"More than"*** parameter type, which means the null hypothesis is rejected if the test statistic is too large.

Two-tailed tests [2]

As the name suggests, in such tests, the rejection region is seen on both ends of the normal distribution curve. A two-tailed test will test the statistical value in both directions.

Which testing approach is the best?

See Fig. 4.3.

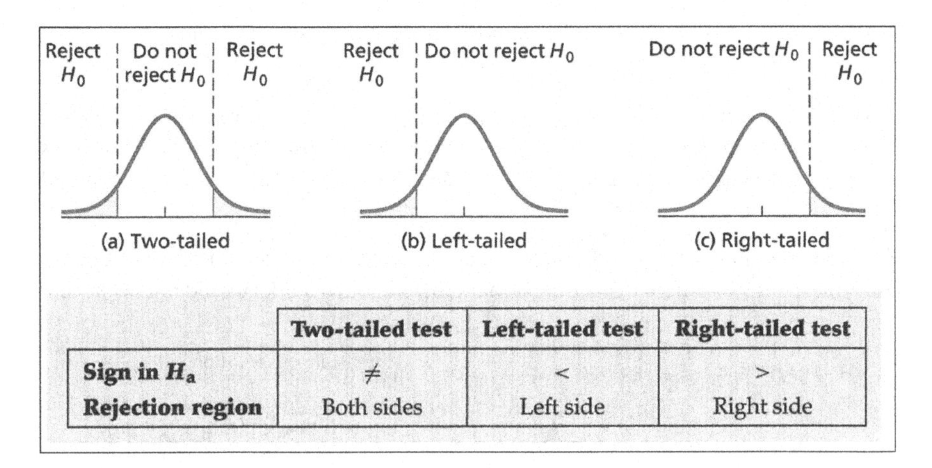

	Two-tailed test	Left-tailed test	Right-tailed test
Sign in H_a	$\neq$	$<$	$>$
Rejection region	Both sides	Left side	Right side

Fig. 4.3 One and two-tailed tests

The answer is two-tailed tests

This is simply because it takes into consideration both positive and negative impacts. When testing has to be done for forecasting and risk mitigation, two-tailed test results are vital as they report valuable, unbiased insights more confidently by double checking the test statistic value in both directions. One drawback with this testing procedure is, it is time consuming and sometimes expensive too.

6. **The p-value** [3]

A p-value, or probability value, is a number describing how likely it is that your data would have occurred by random chance. This denotes the amount of risk that the researcher has to take when rejecting the null hypothesis. The shaded region in Fig. 4.4 represents the *p*-value.

Example 4 Consider a test where the null hypothesis states that, "The average marks of senior students in mathematics at university is 70".

Suppose 80 seniors are selected at random and the sample mean of their scores in mathematics is 72. The researcher wants to know *P* (sample mean $\geq$ 72 given the true value = 70). Using an appropriate test statistic, we compute the *p*-value and it turns out to be **0.08**. This means that if the experiment is repeated 100 times, then 8 times we can expect to see a sample mean mark to be greater than 72.

Example 5 Consider a study for evaluating the effectiveness of a new method of teaching. Group A studied with the new method and secured an average of 70 marks. Group B did not study with the new method and secured an average score of 60 marks.

The researcher starts the study by assuming that the difference of 10 marks between the groups is only due to chance, in other words, the new method does not affect the scores of students. Let us assume, that the results yield a *p*-value of 0.05.

Meaning: Assuming the two groups of students were the same from the start of the study, there's a 95% chance that the 10-mark difference in marks would not be present

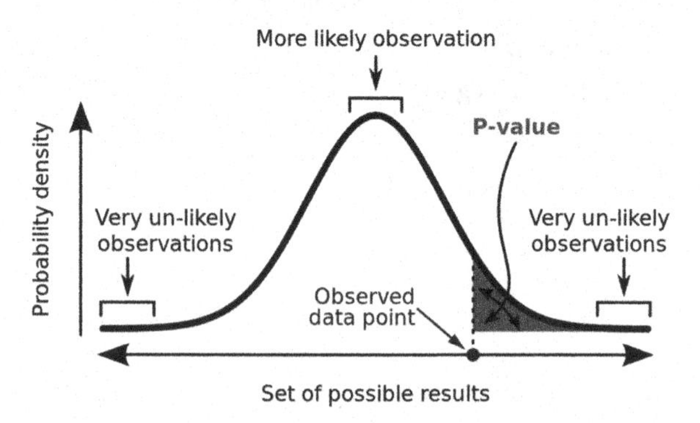

Fig. 4.4 The *p*-value

if the new method of teaching had no effect. The researcher would thus conclude that the new method is effective.

What is the meaning of "rejecting null hypothesis"?

In reality, the null hypothesis is never accepted. We either reject it or fail to reject it. This is simply because you can't prove a negative. If we don't reject the null hypothesis, we don't conclude that it's true. **We just say that the null hypothesis is a possibility**.

Are critical region and p-values the same?

- The critical value is related to test statistic, whereas the significance level is related to p-value.
- The critical value is the boundary of the rejection region. The significance level determines the critical value, and therefore the rejection region, and vice versa.
- Both should be determined prior to data collection.
- The test statistic and the p-value are calculated after collecting the data under the assumption the null hypothesis is true.
- Different samples will result in different test statistic and p-values, but the rejection region and significance level will not change.

Thus, the decision rule can be in two ways

1. Reject H_0 if the test statistic is in the rejection region.
2. Reject H_0 if the p-value is less than the significance level.

Are alpha values the same as p-values?

Alpha α is the value against which p-values are measured and it tells how extreme observed results must be so that the null hypothesis is rejected. This value is related to the confidence level of the test. In general, for results with a $k\%$ level of confidence, the value of alpha is $1 - (k/100)$.

Example 6 If confidence level is 95% level,

$$\alpha = 1 - 0.95 = \mathbf{0.05}.$$

If confidence level is 90% level,

$$\alpha = 1 - 0.90 = \mathbf{0.10}.$$

Why is α at 5% in most of the cases?

A significance level of 0.05 indicates a **5% risk of concluding that a difference exists when there is no actual difference**. Most researchers have agreed that confidence of 95% is appropriate. Historically too, this value has been accepted as standard and α gives the probability of type I error.

Fig. 4.5 Meaning of significance

What is the meaning of "statistically significant"? How do we determine this?

- If the **p-value** is less than or equal to alpha, the null hypothesis is rejected and the result is considered statistically significant which means that it is **not due to chance alone**. There must be some corrections and checks in the process.
- The **p-value** is greater than alpha, we fail to reject the null hypothesis and say that the result is not statistically significant which means we are reasonably sure that our observed **data can be explained by chance alone**.

Note: The smaller the value of alpha, the more difficult it is to claim that a result is statistically significant and the larger the value of alpha, the easier it is to claim that a result is statistically significant.

A Note on Standard Error

Definition
The standard deviation of means of the sample is called the standard error. In inferential statistics, we draw a sample from the population and infer the characteristics of population based on sample. It is very evident that there will be a difference in the means of sample and population, however representative the sample drawn is. This difference is termed as the standard error and is calculated as the ratio between standard deviation and the square root of n samples. $\text{S.E} = \frac{\sigma}{\sqrt{n}}$.
Interpreting standard error:

- As the standard error decreases, the sample becomes a better representative of the population.
- The standard error reveals the accuracy of the sample means when compared to the population mean. If standard error values are high, that means the data is highly scattered and it is an inaccurate representation of the true population mean (Fig. 4.5).

Steps involved in hypothesis testing: Hypothesis testing is conducted systematically with predefined procedures. These steps ensure that the test solves the purpose of study. The sequential steps to be followed are:

Approaches to hypothesis testing: There are three defined approaches in hypothesis testing based on the confidence intervals, critical values and the *p*-values. All these three methods differ only in terms of how the problem statement is defined. Solution wise, all the three approaches work well in their own ways. Among the three, the *p*-value approach is the most popularly used approach.

1. **The test statistic approach**: Here, we compute a test statistic from empirical data and then compare it with a critical value. If the test statistic is larger than the critical value or if the test statistic falls into the rejection region, the null hypothesis is rejected.
2. **p-value approach**: Here, we compute the *p*-value based on a test statistic and then compare it with the significance level (test size). If the *p*-value is smaller than the significance level, we reject the null hypothesis.
3. **Confidence interval approach**: Here, construct the confidence interval and examine if a hypothesized value falls into the interval (Tables 4.1 and 4.2).

The null hypothesis is rejected if the hypothesized value does not fall within the confidence interval.

Table 4.1 Interpretation of *p*-value

p-Value	Evidence against H_0
$p > 0.10$	Weak or no evidence
$0.05 < p \leq 0.10$	Moderate evidence
$0.01 < p \leq 0.05$	Strong evidence
$p \leq 0.01$	Very strong evidence

Table 4.2 Detailed steps of hypothesis testing in all 3 approaches

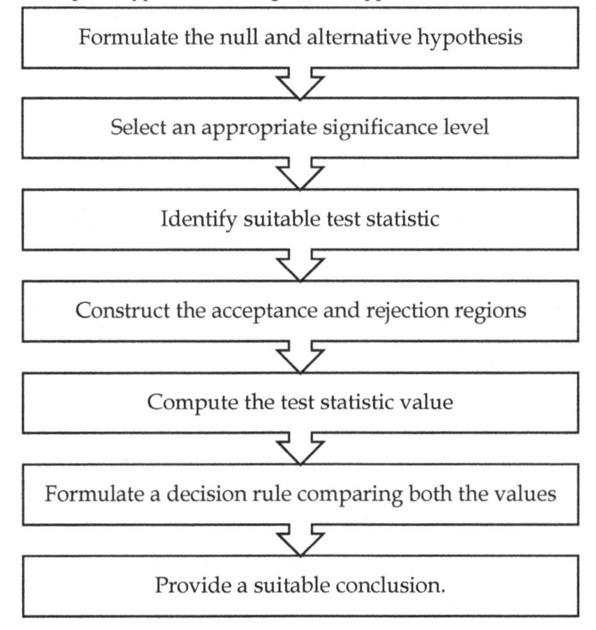

Take a note of how these real-life situations can all be translated into the language of statistical tests of hypotheses [4].

Scenario 1
When an engineer has to decide based on sample data whether the average life of a certain tyre is at least 20,000 miles.

Solution
We might say that the engineer has to test the hypothesis that Θ, the parameter of an exponential distribution is at least 20,000 miles.

Scenario 2
When an agronomist has to decide based on experiments whether one kind of fertilizer produces a high yield for soybeans.

Solution
We might say that the agronomist has to decide whether $\mu_1 > \mu_2$ where μ_1 and μ_2 are the means of two normal populations.

Scenario 3
When a manufacturer of pharmaceutical products has to decide based on samples whether 99% of all patients given a new medication will recover from a certain disease.

Solution
We might say that the manufacturer has to decide whether p, the parameter of binomial distribution equals 0.99.

Errors and Types of Errors

Steps	Test statistic approach	p-value approach	Confidence interval approach
Step 1	State the Null and alternative hypothesis	State the null and alternative hypothesis	State the null and alternative hypothesis
Step 2	Determine the test size alpha, α and find the critical value (C.V.)	Determine a test size alpha, α	Determine a test size alpha α, a sample value and value of standard normal variate at the value of α
Step 3	Compute the test statistic (T.S.)	Compute a test statistic (T.S.) and its p-value	Construct the $(1-\alpha)100\%$ confidence interval
Step 4	Reject the null hypothesis if T.S. > C.V.	Reject the null hypothesis if $p < \alpha$ (alpha)	Reject the null hypothesis if a hypothesised value does not exist in the confidence interval

(continued)

(continued)

Steps	Test statistic approach	*p*-value approach	Confidence interval approach
Step 5	Give a substantiative interpretation	Give a substantiative interpretation	Give a substantiative interpretation

In analytics, we either **accept or reject** a hypothesis, and in reality, the same hypothesis can either be **true or false**. Errors are committed when the truth is rejected and false is accepted. Therefore, these are the two types of errors.

Type I error

The rejection of the null hypothesis when it is true is called **type I error**. The probability of committing a type I error is denoted by **Alpha – α**.

Type II error

The acceptance of the null hypothesis when it is false is known as type II error. The probability of committing a type II error is denoted by β (Fig. 4.6).

A popular saying of the two types of errors are [5].

Type I Error: An alarm without a fire.

Type II Error: A fire without an alarm.

Which is more important for the researcher—Type I or Type II errors? Why?

Usually, both errors lead to wrong results. It is very important to take note of the way in which the hypothesis is formulated while checking for such errors (Fig. 4.7).

The criminal justice system prefers type II errors to type I errors. The best example is when a defendant does not necessarily have to prove that they are innocent of the crime they have been charged with, but just has to raise enough reasonable doubt that prevents the court from declaring that they committed the crime.

Type I—Error is based on false-positive results.

Type II—Error is based on false-negative results. Also called **error of omission**.

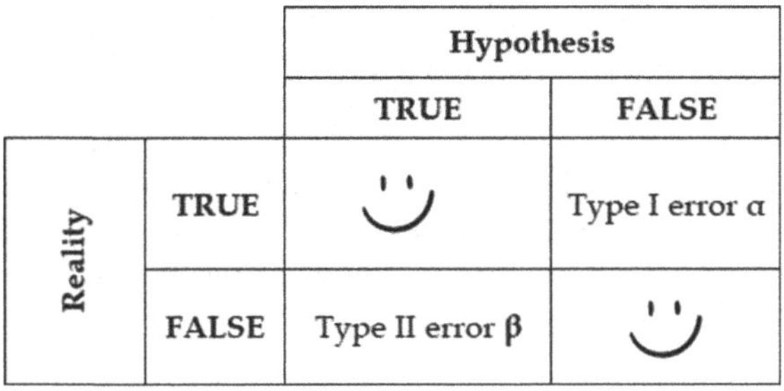

		Hypothesis	
		TRUE	FALSE
Reality	TRUE	☺	Type I error α
	FALSE	Type II error β	☺

Fig. 4.6 Type I and type II errors

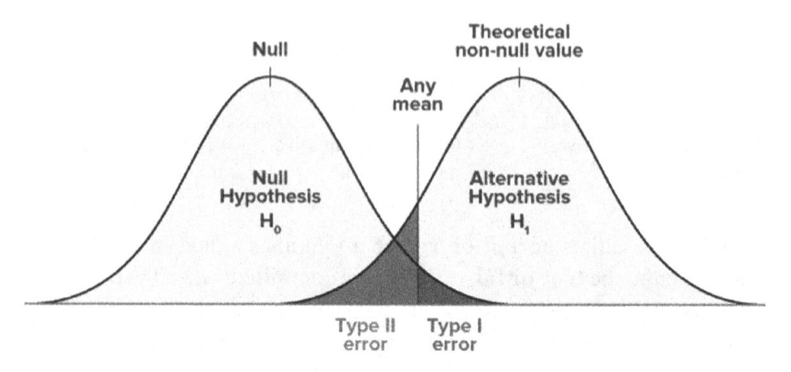

Fig. 4.7 Type I and Type II errors illustrated using a normal curve

Example 7 Consider an epidemic study on a deadly disease, reporting test results as negative when the patient is actually infected is an example of type II error. In this case, type II error is more important than type I error.

Example 8 A person is found guilty of a crime he didn't commit and is unfairly sentenced to death. In such a case type I error is more important than type II. The basic objective is that an innocent man should not be convicted and letting go of a guilty person is better, though not correct.

Case Study [6]

A Wall Street Journal investigative reporter broke the news of a scandal about a medical device maker, Boston Scientific, which reported to the FDA that a new device was better than a competing device. Their conclusions were based upon the results from a randomized trial in which the significance test showing the superiority of their device had a *p*-value of 0.049, just under the criterion of 0.05 that the FDA used for statistical significance. However, the reporter found that the *p*-value was not significant when calculated using 16 other test procedures that he tried. The *p*-values from those procedures averaged 0.051. According to the news story, that small difference between the reported *p*-value of 0.049 and the journalist's recalculated *p*-value of 0.051 was "**the difference between success and failure**".

Discussion: It is not correct to classify the success or failure of this new device according to whether or not the *p*-value falls barely on one side or the other of an arbitrary line, especially when the discussion revolves around the third decimal place of the *p*-value. No sensible interpretation of the data from the study should be affected by the news in this newspaper report.

Power of a test

This is the probability/ability in rejecting the null hypothesis when the null hypothesis is false. It is given by 1 – β. Power is the probability of avoiding a type II error.

$$\begin{aligned} &P(\text{not making a type II error}) \\ &\quad = 1 - P(\text{accepting Ho when it is false}) \\ &\quad = P(\text{Rejecting Ho when it is false}) = P(\beta) \end{aligned}$$

Concept of "degrees of freedom"

The degrees of freedom ***degrees of freedom*** in statistics indicate the number of logically independent values that can vary in an analysis without breaking any constraints.

To explain in simple terms: Consider this set of clothes shown in Fig. 4.8, you had a choice to choose one every day. Then, on the

Day 1: We have 5 options to choose from,
Day 2: We have 4 options,
Day 3: We are left with 3 options and

Fig. 4.8 Choice of clothes

Day 4: We have just 2 options. After Day 4, you had to select the one that was remaining without any choice. So, you were given the freedom to choose till day 4, which is **"n − 1"**.

The same concept is applicable for the variables in testing. In each test case, there are different degrees of freedom that are assigned to variables which will be explained as and when studied.

4.2 Parametric and Non-parametric Tests

A **parametric** test is a statistical test that makes assumptions about the parameters of the population distribution(s) from which the data is drawn.

Case I
To test the property of a single population. For example,

1. Testing the mean/median value.
2. The percentage or proportion of the population which has a particular property.
3. The dispersion/spread of a specific property in the population (Table 4.3).

Different types of parametric test

Let's take a look at the popular parametric tests that are used for

- Comparing two distributions.
- Identifying the relationship between parameters and their values.

These tests can measure the mean, variance and analyses the proportions. The tests performed vary based on the data type which is dependent or independent data, or based on the number of samples such as 1, 2 or 3 or more samples (Fig. 4.9).

T1. One-Sample Testing of Means

Let's begin with understanding the hypothesis testing of mean in a single sample. There are two testing methods which are the *z*- and *t*-test.

Table 4.3 Measurements and type of tests

Type of measurement	Type of test
Interval/ ratio	Parametric
Nominal/ordinal	Non-parametric
Ordinal scale with ratings	Parametric

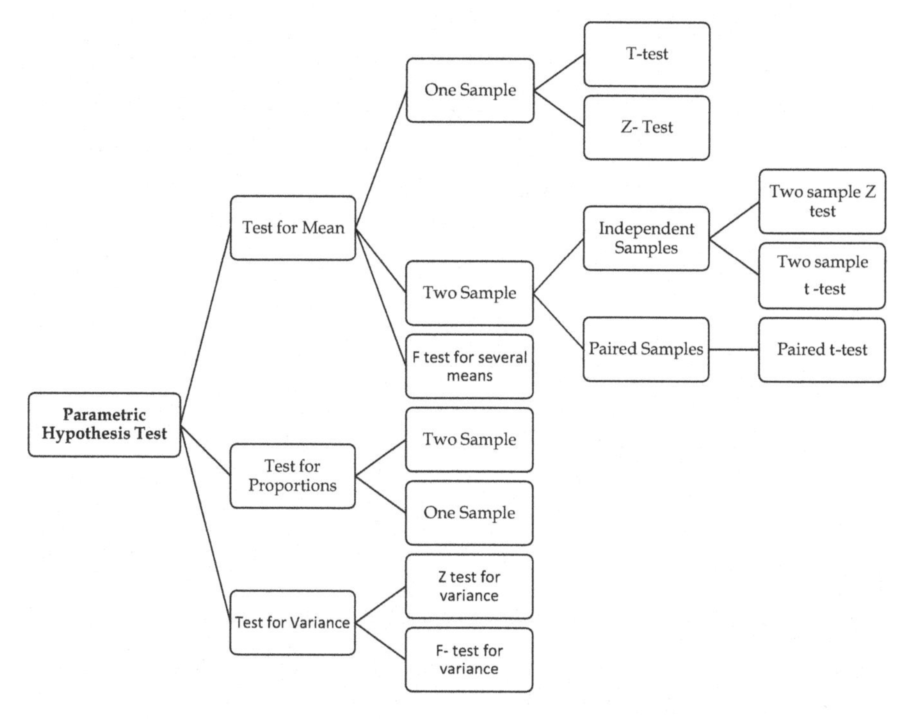

Fig. 4.9 Different types of parametric tests

T.1.1. Testing the Mean Value Z-Test

Assumptions

- Sample size > 30.
- Population mean and variance are known.
- The data follows the normal distribution.

The one-sample statistical *z*-test test is suitable for testing whether the sample means $\overline{x}$ is different from the population mean μ. In this test, we are aware of the population variance value.

Null hypothesis $\mathbf{H_0}$: The population mean and sample mean are not significantly different.

$$\mu = \mu_0$$

Alternate hypothesis $\mathbf{H_1}$: The population mean and sample mean are significantly different. It can either be, $\mu > \mu_0$, $\mu < \mu_0$ for a one-tailed test. For a two-tailed test, it is $\mu \neq \mu_0$.

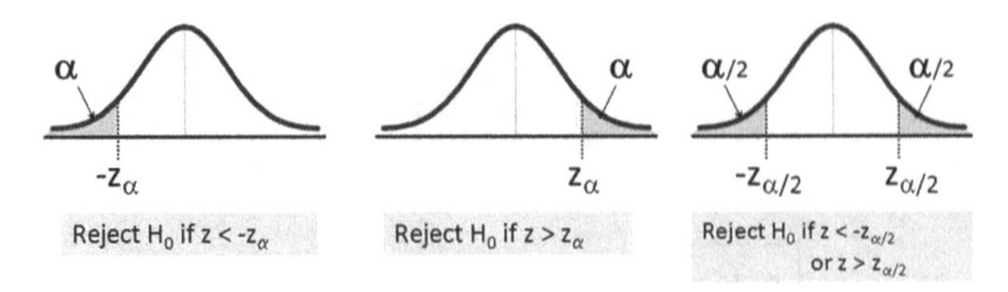

Fig. 4.10 Graphical representation of the decision rule

Table 4.4 Decision rule for single mean Z test

Alternative hypothesis	Decision rule
$\mu < \mu_0$	$Z < -Z_\alpha$
$\mu > \mu_0$	$Z > -Z_\alpha$
$\mu \neq \mu_0$	$-Z_{\alpha/2} > Z > Z_{\alpha/2}$

The test statistic: $Z = \frac{\overline{x}-\mu}{\frac{\sigma}{\sqrt{n}}} \sim N(0, 1)$.

Note: For two-tailed test, testing is done at $\alpha/2$ level of significance.

Decision Rule: The decision rule used for rejecting the null hypothesis in favour of the alternative is given as follows (Fig. 4.10; Table 4.4).

Standard Error: Estimated standard error for the mean difference between the sample mean μ and the population mean value μ_0 is, $S = \frac{\sigma}{\sqrt{n}}$.

Example 9 A sample of 50 students is taken from a large population, and the mean height of the students is found to be 64 in. with a population standard deviation of 4 in. Can it be reasonably regarded that the mean height of the students of the population is 64 in.?

Suitable Tool: Test for a single mean when variance is known.

Reason: There is a single sample and the population SD is known. The sample is fairly large. (Size is 50). So, we use the Z-test for testing a single mean.

Null hypothesis H$_0$: The mean height of the students of the population is 66 in. $\mu = 66$ in.

Alternative hypothesis H$_1$: The mean height of the students of the population is not equal to 66 in. $\mu \neq 66$.

Test Statistic: $Z = \frac{\overline{x}-\mu}{\frac{\sigma}{\sqrt{n}}} \sim N(0, 1)$.

Level of Significance: Alpha, $\alpha = 0.05$. Two-tailed tests, $\alpha/2 = 0.025$.

Decision Rule: Reject H$_0$ if $|\mathbf{Z}| \geq \mathbf{Z}_{\frac{\alpha}{2}}$.

Calculation:

Given **n** $= 50, \overline{x} = 64, \sigma = 4, \mu_0 = 66, \alpha = 0.05$

$$Z = \frac{\overline{x} - \mu}{\frac{\sigma}{\sqrt{n}}} \sim N(0, 1) = \mathbf{-3.53}$$

$$Z_{\frac{\alpha}{2}=1.96}$$

Inference: Since $|-3.53| > 1.96$ we reject the null hypothesis.

Conclusion: The mean population height of the students is not equal to 66 in.

T.1.2 One-Sample Testing of Means: t-Test

Assumptions

- Population mean is known but the variance is unknown.
- Sample size < 30.
- Data is normally distributed.

This test is similar to that of the single sample Z-test. The only difference is when the sample size is small, $n < 30$, and when variance is unknown, t-tests are an appropriate statistical test. This test uses sample standard deviation $\boldsymbol{s}$ to estimate population SD which is $\boldsymbol{\sigma}$.

Null hypothesis H$_0$: The population mean and sample mean are not significantly different.

$$\mu = \mu_0$$

Alternative hypothesis H$_1$: The population mean and sample mean are significantly different. It can either be $\mu > \mu_0$, $\mu < \mu_0$ for a one-tailed test. For a two-tailed test, it is $\mu \neq \mu_0$.

The test statistic

The test statistic used for testing the null hypothesis is given by

$$t = \frac{\overline{x} - \mu_0}{\frac{s}{\sqrt{n}}} \sim t_{n-1} \quad \text{where, } s \sqrt{\frac{1}{n-1}\left(\sum_{i=1}^{n} x_i^2 - n\overline{x}^2\right)}$$

$\overline{x}$ is the sample mean, s is the sample standard deviation and n is the sample size. The degree of freedom is $n - 1$.

Standard Error: Standard error = $\frac{s}{\sqrt{n}}$.

Decision Rule: The decision rule used for rejecting the null hypothesis in favour of the alternative is given as follows (Table 4.5).

Case Study

Certain pesticides are packed into bags by a machine. A random sample of 10 bags is drawn and their contents are found to weigh (in kilograms) as follows.

50 49 52 44 45 48 46 45 49 45

Test if the average packing can be taken as 50 kg.

Suitable Tool: Test for a single mean when variance is unknown.

Reason: σ is unknown and has to be replaced by a quantity that can be derived from the sample.

Null hypothesis: $\mathbf{H_0}$ The average weight of the packing is 50 kg $\mu = 50$.

Alternative hypothesis: $\mathbf{H_1}$ The average weight of the packing is not equal to 50 kg, $\mu \neq 50$.

Test Statistic: $\frac{\overline{x}-\mu_0}{\frac{s}{\sqrt{n}}} \sim t_{n-1}$

Level of Significance: $\alpha = 0.05$.

Decision Rule: Reject if $|t| \geq t_{\frac{\alpha}{2},n-1}$

Calculation:

Given $n = 10$, $\overline{x} = 47.3$, $\mu_0 = 50$, $\alpha = 0.05$

$$\sqrt{\frac{1}{n-1}\left(\sum_{i=1}^{n} x_i^2 - n\overline{x}^2\right)} = \mathbf{2.669}$$

$$t = \frac{\overline{x} - \mu_0}{\frac{s}{\sqrt{n}}} = \mathbf{-3.199}$$

$$|t| = 3.199 \text{ and } t_{\frac{\alpha}{2},n-1} = \mathbf{2.262}$$

Since 3.199 > 2.262 we reject the null hypothesis.

Table 4.5 Decision rule for single means mean t-test

Alternative hypothesis	Decision rule
$\mu < \mu_0$	$t < -t_{\alpha,n-1}$
$\mu > \mu_0$	$t > t_{\alpha,n-1}$
$\mu \neq \mu_0$	$\|t\| \geq t_{\frac{\alpha}{2},n-1}$

Conclusion: The average weight of the packing is not equal to 50 kg.

2. Testing two samples for differences in their means. Here again, there are two testing methods, Z and the t-test. In these tests we test the relationship between sample means of two groups of data.

2.1 Testing Two Sample Means Using Z-test

Testing difference of mean between two large independent samples.

Assumptions

- Two uncorrelated independent samples.
- Sample size > 30.
- Population variances or SD are known.
- Normally distributed data.

Procedure

1. **Null hypothesis: $\mathbf{H_0}$** $= \overline{x_1} = \overline{x_2}$
2. **Alternative hypothesis**: $\mathbf{H_1} = \overline{x_1} \neq \overline{x_2}$ or $\overline{x_1} > \overline{x_2}$ or $\overline{x_1} < \overline{x_2}$
3. **Test statistic**:
   Compute $Z = \frac{\overline{x_1}-\overline{x_2}-0}{\sigma_{\text{DM}}} = \frac{\overline{x_1}-\overline{x_2}}{\sqrt{\frac{\sigma_1{}^2}{N_1}+\frac{\sigma_2{}^2}{N_2}}} \sim N(0, 1)$
   Standard deviation for difference of means is given as
   $\sigma_{\text{DM}} = \sigma\sqrt{\frac{1}{n_1} + \frac{1}{n_2}}$ when SD is common for populations.
   $\sigma_{\text{DM}} = \sigma\sqrt{\frac{\sigma_1{}^2}{n_1} + \frac{\sigma_2{}^2}{n_2}}$ when the SD is different for the populations.
4. Obtain the value of Z_α from the table, where α is the level of significance.
5. Compute $\frac{|\overline{x_1}-\overline{x_2}|-0}{\sigma_{\text{DM}}}$ for a two-tailed test and $\frac{\overline{x_1}-\overline{x_2}-0}{\sigma_{\text{DM}}}$ for a one-tailed test.

Decision Rule

If z falls outside the critical region, accept H_0 or else we state that we do not have enough evidence to reject H_0 (Table 4.6).

Standard error: $\text{S.E} = \sqrt{\frac{\sigma_1{}^2}{N_1} + \frac{\sigma_2{}^2}{N_2}}$.

Example 10 The average IQ of students in a college is 105 with a SD of 8. A random sample of 50 students picked up from the school population shows the average IQ as 108.

Table 4.6 Decision rule for differences of two means mean Z-test

Alternative hypothesis	Decision rule
$\mu < \mu_0$	$Z < --Z_\alpha$
$\mu > \mu_0$	$Z > --Z_\alpha$
$\mu \neq \mu_0$	$-Z_{\alpha/2} > Z > Z_{\alpha/2}$

A. Does the sample mean differs positively from the mean of the school population at 0.01 level of significance?
B. A second sample of size 75 drawn from the same college shows that the mean IQ is 106. Is there a significant difference between the two samples at a 0.05 level of significance?

Suitable Tool: Test for the equality of two means.

Reason: There are two samples and their standard deviations are known. We need to test for differences between means.

Solution

A. **Given**: $\mu = 105, \sigma = 8, \overline{x_1} = 108, \overline{x_2} = 106, n_1 = 50, n_2 = 75.$

Null hypothesis: $\mathbf{H_0}$: $\mu = \mu_0$.

Alternative hypothesis: $\mathbf{H_1}$: $\mu \neq \mu_0$.
Here $\alpha = 0.01$. We use the right-tailed z-test. We also have $Z_{0.01} = 2.327$. $\sigma_{\overline{x_1}} = \frac{\sigma}{\sqrt{n_1}} = 1.131$ and

$$Z = \frac{\overline{x_1} - \mu}{\sigma_{\overline{x_1}}} = 2.652$$

Conclusion: The obtained value of z is > 2.327. Therefore, z is in the critical region hence we reject the null hypothesis. The sample drawn does not represent the population from which it is drawn.

B. The two means are uncorrelated since the two samples are drawn independently from the same population. The two-tailed z-test is most suitable. The common SD shared by the two samples is SD = 8

$$\sigma_{\text{DM}} = \sigma\sqrt{\frac{1}{n_1} + \frac{1}{n_2}} = \mathbf{0.461}\ \mathbf{Z} = \frac{|\overline{x_1} - \overline{x_2}| - 0}{\sigma_{\text{DM}}} = \mathbf{1.369}$$

Critical region: $Z \geq 1.96$ at 0.05 level.

Conclusion: We accept the null hypothesis since z lies outside the critical region. Therefore, the two samples have been drawn from the same population.

2.2. Testing Two Sample Means Using t-Test

Testing difference of mean between two small independent samples.

Assumptions

- Two uncorrelated independent samples.
- Sample size < 30. (Sometimes applicable even when size is big and variance is unknown.)
- The unknown variances of the 2 populations are equal. Normally distributed data.

Pooled Variance

This is an unbiased estimator of two combined variances obtained from two or more samples. It is also a weighted average of two standard deviations. Denoted as $\mathbf{S_p}^2$, and it is calculated as

$$S_p = \sqrt{\frac{\sum(x_1 - \overline{x_1})^2 + \sum(x_2 - \overline{x_2})^2}{n_1 + n_2 - 2}}$$

Null hypothesis: $\mathbf{H_0}$: $\mu_1 = \mu_2$.

Alternative hypothesis: $\mathbf{H_1}$: $\mu_1 \neq \mu_2$.

Test statistic: $t == \frac{|\overline{x_1} - \overline{x_2}|}{\sigma_{\text{DM}}} \sim t_{n_1, n_2 - 2}$

where $\sigma_{\text{DM}} = S_p\sqrt{\frac{1}{n_1} + \frac{1}{n_2}}$ and $S_p = \sqrt{\frac{\sum(x_1 - \overline{x_1})^2 + \sum(x_2 - \overline{x_2})^2}{n_1 + n_2 - 2}}$.

Degrees of freedom: $n_1 + n_2 - 2$.

Level of significance: $\alpha = 5\%$.

Decision rule: Here **v** refers to the degrees of freedom which is $\mathbf{n_1 + n_2 - 2}$ (Table 4.7).

Example 11 In constructing a building, a person buys cement bags of the same brand from the same shop. In the first lot of 100 bags, the average weight of the bags was 0.8 kg less than the prescribed 50 kg with SD = 0.8 kg. The second lot of 150 bags shows a loss of 1.5 kg per bag with SD = 1.3 kg. Does the data indicates that the consignments were from the lot at a 0.05 level of significance?

Suitable Tool: Test for the equality of means when the population variance is unknown.

Reason: There are two samples and their means have to be compared.

Solution
Given: $\overline{x}_1 = 0.8$, $\overline{x}_2 = 1.5$, $\sigma_1 = 0.8$, $\sigma_2 = 1.3$, $n_2 = 150$, $n_1 = 100$.

Table 4.7 Decision rule for differences of two means mean *t*-test

Alternative hypothesis	Decision rule
$\mu < \mu_0$	$T < t_{\alpha,v}$
$\mu > \mu_0$	$T > < t_{1-\alpha,v}$
$\mu \neq \mu_0$	$\|T\| > t_{1-\alpha,/2,v}$

Null hypothesis: $\mathbf{H_0}$: $\mu_1 = \mu_2$.

Alternative hypothesis: $\mathbf{H_1}$: $\mu_1 \neq \mu_2$.

The SD is unknown, and hence, we use the t-test for studying the significance of the difference between the two means. A two-tailed t-test at $\alpha = 0.01$ level of significance is used here.

$$\sigma_{\mathrm{DM}} = \sqrt{\frac{\sigma_1^2}{n_1} + \frac{\sigma_2^2}{n_2}} = 0.146$$

$$t == \frac{|\overline{x_1} - \overline{x_2}| - 0}{\sigma_{\mathrm{DM}}} \sim t_{n_1,n_2-2} = \mathbf{4.705}$$

Degrees of freedom d.f: $n_1 + n_2 - 2. = 248$.

Critical value for $t_{0.01,248} = \mathbf{2.595}$.

The calculated value of t which is 4.705 is highly significant as it is greater than the critical value of t, which is 2.595. Hence, the null hypothesis is rejected in favour of the alternative hypothesis which says that the two consignments come from different lots.

Randomness, normality, SD of populations unknown but assumed to be equal, uncorrelated small samples—student's t ratio. When there are two independent samples of equal sizes (say 30 or less), the standard error of the difference between the means depends upon the calculated SDs by the formula SD$= \sqrt{\sum \frac{x-\bar{x}^2}{n-1}}$ and the number of degrees of freedom is $n_1 + n_2 - 2$. The appropriate test to be used when testing the difference between sample means when the sample sizes are small is the student's—t ratio.

Case Study

The amount of a certain raw material used to produce linoleum is a critical factor in linoleum durability. A researcher for linoleum manufacturers believes that the mean concentration of the chemical is different in the raw material obtained from two suppliers. To find whether or not those beliefs can be supported by objective data, the researcher takes random samples from the raw materials provided by the two suppliers and determines the concentration of the chemical in each specimen. The results are as follows:

Supplier 1

60.9 49.8 95.3 40.6 51.6 69.7 58.0 59.6 47.8 46.9 47.6 67.3

Supplier 2

72.9 67.3 81.4 89.4 86.5 51.1 72.9 74.0 77.8 86.4 82.0 77.6 74.8 50.7
61.0 57.4 61.0 57.8 74.7 89.4

Suitable tool: Test for the equality of means when population variances are unknown but equal.

Reason: There are two samples, and nothing is mentioned about their variances, but the sample variance can be found from the data given.

Solution
Null hypothesis H_0: The mean concentration of the chemical is equal to the raw material obtained from the 2 suppliers $\mathbf{H_0 = \mu_1 = \mu_0}$.

Alternative hypothesis: The mean concentration of the chemical is not equal in the raw material obtained from the 2 suppliers $\mathbf{H1 = \mu_1 \neq \mu_0}$.

Test statistic: $t = \frac{|\overline{x_1} - \overline{x_2}|}{\sigma_{DM}} \sim t_{n_1, n_2 - 2}$.

$\sigma_{DM} = S_p \sqrt{\frac{1}{n_1} + \frac{1}{n_2}}$ and $S_p = \sqrt{\frac{\sum (x_1 - \overline{x_1})^2 + \sum (x_2 - \overline{x_2})^2}{n_1 + n_2 - 2}}$ is the pooled variance.

Level of significance: $\alpha = 0.05$.

Decision rule: Reject $\mathbf{H_0}$ if $|t| \geq t_{\alpha/2, n_1, n_2 - 2}$

Calculations: Given the data above, we get the following: $\overline{x}_1 = 55.425, \overline{x}_2 = 72.305$, $n_1 = 12$, $n_2 = 20$.

$$S_p = \sqrt{\frac{\sum (x_1 - \overline{x_1})^2 + \sum (x_2 - \overline{x_2})^2}{n_1 + n_2 - 2}} = 127.3, t = \frac{|\overline{x_1} - \overline{x_2}|}{\sigma_{DM}} = |-4.09| = 4.09$$

$$t_{\alpha/2, n_1, n_2 - 2} = 2.045$$

Inference: since 4.09 > 2.045, we reject $\mathbf{H_0}$.

Conclusion: The mean concentration of the chemical is not equal in the raw material obtained from the 2 suppliers.

T2.3 Paired-samples test

Usage: Just like the *t*-test but when the researcher can arrange the two sets in pairs.
Why do we use the paired samples test?

Because by pairing the treatments we can remove random variability from the *t*-test.

Important: Pairing should not be done arbitrarily.

Paired *t*-test assumptions

1. The data is continuous (not discrete).
2. The differences **d** for the matched pairs follow a normal probability distribution.
3. The sample of pairs is a simple random sample from its population. Each individual in the population has an equal probability of being selected in the sample (Fig. 4.11).

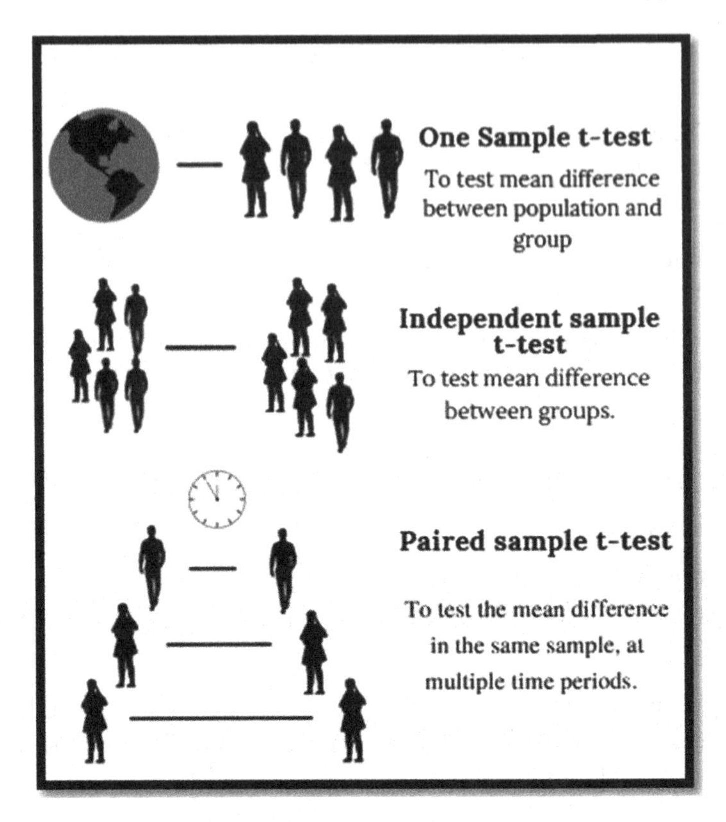

Fig. 4.11 Illustrating different types of *t*-tests

Note: The critical ratio student's *t*-test is appropriate in testing the difference between sample means when SD in the population is unknown. When the sample is 100 or more, the *Z*-test gives almost the same results as the *t*-test. Testing difference of mean from large samples, when means are correlated—student's *t*-test is used.

T2.4 Snedecor's F-test for comparison of several means

Snedecor's *F*-test is most suitable for the simultaneous comparison of several means.

Procedure

Null hypothesis

$H_0 \quad \mu_0 = \mu_1 = \mu_3 = \cdots = \mu_k.$

Alternate hypothesis: H_1: $\mu_0 \neq \mu_1 \neq \mu_3 \neq \cdots \neq \mu_k.$

Calculations: Compute mean and SD for each sample.

Compute $\sum_k (\overline{x_i} - \overline{x})$ where $\overline{x_i}$ the mean of the *i*th sample is and $\overline{x}$ is the combined mean of all samples. The above expression can be rewritten as $\sum_k (\overline{x_i} - \overline{x}) = \frac{\sum_k \overline{x_i}^2 - (\sum_k \overline{x_i})^2}{k}$.

Compute $\sigma^2 = \sum \frac{S_i{}^2}{k}$ where the unbiased estimate common population variance is and n is the common size for all samples and $F = \frac{n \sum \overline{x_i} - \overline{x}}{(k-1)\sigma^2}$.

Degrees of freedom: $df_1 = k - 1$ and $df_2 = k(n - 1)$.

Choose the level of significance, α and obtain F_α for df_1 and df_2 from the F distribution table.

Decision rule

The critical region is at α level of significance.

If $F \geq F_\alpha$ reject H_O

If $F < F_\alpha$ accept H_O.

Case Study

Five plants each of the same height and age and each of four varieties of Chrysanthemum are selected at random and planted in equal-sized pots. The soil mix of equal quantity is used in each pot, and all the pots received similar attention. Below are the observed average height and variance after 6 weeks of the plantation. Does the data shows that the growth patterns of different varieties of plants are different? (Table 4.8).

Suitable Tool: Snedecor's F-test is a suitable tool.

Reason: Small samples whose variances are unknown.

Solution

Null hypothesis: All the sample means are equal. $H_{0:}\ \overline{x}_0 = \overline{x}_1 = \overline{x}_2 = \cdots = \overline{x}_k$.

Alternative hypothesis: All the sample means are not equal. $H_{1:}\ \overline{x}_0 \neq \overline{x}_1 \neq \overline{x}_2 \neq \cdots \neq \overline{x}_k$.

For easy computation, the sample means are reduced by 30 cm so that the new means now read as 2, – 2, 0 and 4.

$$\frac{\sum_k \overline{x_i}^2 - \left(\sum_k \overline{x_i}\right)^2}{k} = 20 \text{ and } \sigma^2 = \sum \frac{S_i^2}{k} = 11.8,$$

$$df_1 = k - 1 = 4 - 1 = 3 \text{ and } df_2 = k(n - 1) = 4 * (5 - 1) = 16$$

Critical value: $F_{0.05} = 3.2389$

Table 4.8 Mean and variance of 4 types of chrysanthemums

Varieties	1	2	3	4
Mean height	32	28	30	34
Variance	10	12.5	8.3	16.4

$$F = \frac{n \sum(\overline{x_i} - \overline{x})}{(k-1)\sigma^2} = 2.8249$$

Conclusion: Since $_F <$ $F_{0.05}$, we accept H_0. The small bit of difference in the average height of 4 groups of plants can be solely explained by random fluctuations.

T.3 Testing the proportion of the population which has a particular property

In this testing method, we will learn about testing population proportions.

- Does a single population proportion equal a particular value?
- Whether there is a difference in the two population proportions? Are they equal?

We'll start our exploration of hypothesis tests by focusing on population proportions. Specifically, we'll derive the methods used for testing:

T3.1 Test for mean of proportions, single mean: Z-test

Assumptions

1. Randomness.
2. Data is normally distributed.
3. Observed values are mutually independent and have the same probability of occurrence.

Caution: The sample size should be such that (n * p) $np > 5$ and p should not be less than 0.10.

Caution: The sample size should be such that $np > 5$ and p should not be less than 0.10.

Procedure

1. Define the null and alternative hypothesis.
2. Select a suitable alpha value.
3. Compute p and see if **np** > 5 and 0.10 < **p** < 0.90.
3. Compute SE (p) or $\sigma_p = \sqrt{\frac{P(1-P)}{n}} \sim N(0,1)$ where P is the proportion in the population.
   $P(1-P) =$ PQ is the population variance.
4. Compute CR $Z = \left|\frac{p-P}{\sigma_p}\right|$ for a two-tailed test or $Z = \frac{p-P}{\sigma_p}$ for a one-tailed test.
5. Read the appropriate value for Z_α from the table.
6. The critical region is $|z| \geq Z_\alpha$ for a two-tailed test and $z \geq Z_{2\alpha}$ for a one-tailed test.
7. Accept H_0 if $z \geq Z_\alpha$.

Case Study

General nucleonics, the major energy corporation, is attempting to take over all Sergeant Nucleonics. General's consultant reports that 60% of Sergeant's shareholders support the takeover bid. To be sure of this, the General's president requests a telephone survey of a random sample of Sergeant's shareholders. The staff polls 1500 shareholders and finds that 784 support the takeover bid. At a level of significance of 0.01, does the poll refute the consultant's report?

Suitable tool: Test for a single proportion.

Reason: The proportion of shareholders is given in the question as also the sample proportion.

Thus, the test for proportion is the right choice.

Solution

Null hypothesis: H_0 The proportion of sergeant shareholders who support the takeover bid is 0.60. $P = 0.6$.

Alternative hypothesis: H_1 The proportion of sergeant shareholders who support the takeover bid is less than 0.60. $P < 0.6$.

Test statistic: $Z = \sqrt{\frac{P(1-P)}{n}} \sim N(0,1)$.

Level of significance: $\alpha = 0.01$.

Decision rule: Reject H_0 if $Z \leq -Z_\alpha$

Calculation

Given: $X = 784$, $n = 1500$, $P = 0.60$

$$P = x/n = 0.52, \sigma_p = \sqrt{\frac{P(1-P)}{n}}$$

$$Z = \frac{p - P}{\sigma_p} = -6.32$$

$$-Z_\alpha = 2.33$$

Inference: Since $-6.32 < -2.33$ we reject the null hypothesis.

Conclusion: The proportion of sergeant shareholders who support the takeover bid is less than 0.60, and hence, the poll refutes the consultant's reports. Testing difference of mean from large samples—when means are uncorrelated, samples are drawn randomly.

T3.2 Test for comparison of 2 proportions: Z-test

To test two samples from different populations on any specific characteristic. The z-test for comparison of 2 sample proportions is the key statistical tool used for assessing if the proportions are significantly different from each other. This test can be used for both large and small samples drawn from the population. For example,

Table 4.9 Decision rule for Z test of two proportion tests

Alternative hypothesis	Decision rule
$p_1 > p_2$	$Z > Z_\alpha$
$p_1 < p_2$	$Z < -Z_\alpha$
$p_1 \neq p_2$	$-Z_{\frac{\alpha}{2}} > Z > Z_{\alpha/2}$

test the mean proportion of wheat harvest and jowar harvest in the year 2014 (year after a severe drought in Rajasthan).

Assumptions

- The sample data collected from both populations is random.
- The population from which samples are drawn follow the normal distribution.
- Samples are independent of each other.
- The standard deviation (σ) of the population is known.
- $np > 5, 0.10 < p < 0.90$.

Null hypothesis $\mathbf{H_0}$: The sample proportions are the same. $\pi_1 = \pi_2$.

Alternative hypothesis $\mathbf{H_1}$: The sample proportions are not the same. $\pi_1 \neq \pi_2$.

Test statistic: Z-test for two proportions $= \frac{\widehat{p_1} - \widehat{p_0}}{\sqrt{p_o - (1 - p_0)\left(\frac{1}{n_1} + \frac{1}{n_2}\right)}}$ where $p_0 = \frac{X_1 + X_2}{n_1 + n_2}$.

where n_1 and n_2 are sample sizes of samples 1 and 2.

x_1 and x_2 are the number of trials, and $\widehat{p_1}$ and $\widehat{p_0}$ are observed proportions of the events in two samples.

Decision rule

See Table 4.9.

Example 12 Customer satisfaction is directly proportional to the number of quality products produced by the company. Thus, a food manufacturer wants to test two of his production lines in the first phase of production. Line A reports 26 defects out of 2050 samples, and Line B reports 15 defects out of 150 samples. At α 5%, is there a significant difference in the defects detected between two production lines? Is it true that Line A produces more defectives than Line B.

Null hypothesis: The defects in both the production lines are the same. $\pi_1 = \pi_2$.

Alternative hypothesis: The defects in both the production lines are not the same. $\pi_1 > \pi_2$.

Test Statistic

Z-test for two proportions = $\frac{\widehat{p_1}-\widehat{p_0}}{\sqrt{p_o-(1-p_0)\left(\frac{1}{n_1}+\frac{1}{n_2}\right)}}$ where

$$p_0 = \frac{X_1 + X_2}{n_1 + n_2}$$

$$\widehat{p_1} = \frac{X_1}{n_1} = \frac{26}{200} = 0.13 \text{ and } \widehat{p_0} = \frac{X_2}{n_2} = \frac{15}{150} = 0.1$$

$$p_0 = \frac{15 + 26}{200 + 150} = \frac{41}{350} = 0.1171$$

$$Z_{\text{cal}} = \frac{0.13 - 0.1}{\sqrt{0.1171(1 - 0.1171)\left(\frac{1}{200} + \frac{1}{150}\right)}}$$

$$= \frac{0.03}{\sqrt{0.1033 * 0.0566}} = \frac{0.03}{0.07649} = 0.3922$$

$$Z_\alpha = 0.6517$$

Decision Rule: Since Z calculated value is less than the Z alpha value. We do not reject the null hypothesis.

T5. Testing the population variance (dispersion/spread)

T4.1 Z-test for population variance

Assumptions

- Data must be selected at random.
- Adheres to the normal distribution.
- Population variance is known.

Procedure

1. Compute standard deviation in the sample.
2. Frame the null and alternative hypothesis based on the statement of the problem. Select the level of significance α. Read from the table of areas under the normal curve z.
3. Compute the standard error of SD by the formula $\sigma_\sigma = \frac{\sigma}{\sqrt{2n}}$
4. Apply Z-test.
5. Find out the critical region.
6. Decision rules to reject/not reject a null hypothesis are similar to the rules of the single mean z-test.

Example 13 The average age at marriage of Indian girls is calculated at 18.6 years with an SD of 1.6 years. A random sample of 25 married girls shows the mean age at marriage is 20.1 years with an SD of 1.3 years. Does the data shows that the variance in the sample is different from the population variance at a 0.01 level of significance?

Appropriate tool: The appropriate test is two-tailed z.

Solution

Given: $\sigma = 1.6$, $S = 1.3$, $n = 25$.

Null hypothesis: The variations of the sample and population are similar. $H_0 : \sigma = S$

Alternative hypothesis: The variations of the sample and population are similar. $H_1 : \sigma \neq S$

$$\sigma_\sigma = \frac{\sigma}{\sqrt{2n}} = 0.2262$$

$$Z = \left| \frac{S - \sigma}{\sigma_\sigma} \right| = 1.326$$

From the normal table, we have $Z_{0.01} = 2.576$.

Conclusion: The calculated value of z falls outside the critical region. Therefore, the null hypothesis is accepted. The sample SD is not statistically different from the population SD.

T4.2 Testing difference in variability between groups

Snedecor's *F*-test for variances

An effective statistical test when we need to test the variances of two or more samples simultaneously.

Assumptions

- Data follows the normal distribution.
- Variance is unknown.
- Sufficiently large sample.
- Samples are collected at random.

Null hypothesis: The variance of samples 1 and 2 is the same. $H_0 : {S_1}^2 = {S_2}^2$.

Alternative hypothesis: The variance of samples 1 and 2 is not the same. $H_1 : {S_1}^2 \neq {S_2}^2$.

Calculation: Compute ${S_1}^2$ and ${S_2}^2$.

Degrees of freedom: Number of samples − 1 which is $N_1 - 1$ and $N_2 - 1$.

Test Statistic: The test statistic is just the ratio of the two variances. Compute $F = \frac{{S_1}^2}{{S_2}^2}$ the larger variance being in the numerator.

Critical Value: Find the value of $P(F)$ at α level of significance from the F table.

Decision rule

See Table 4.10.

Case Study

An attitude survey was conducted on randomly selected MLAs from two states towards the economic policy adopted by the central government.

In state *A*, the Congress Party was in power. And in the other state *B*, a combined Left Front was in power. If $S_A = 10$, $S_B = 15$, $n_A = 20$, $n_B = 25$ can we say the variances differ significantly at 0.05 level of significance?

Suitable Tool: Snedecor's *F*-test.

Reason: We are comparing several variances at the same time.

Solution
Given: $S_A = 10$, $S_B = 15$, $n_A = 20$, $n_B = 25$.

Null hypothesis: The variance in the attitude of Congress party MLAs is equal to the MLA members in the Left Front run state. $\text{H}_0 = {S_A}^2 = {S_B}^2$.

Alternative hypothesis: The variance in the attitude of Congress party MLAs are equal to the M.L.A members in the Left Front run state. $\text{H}_0 = {S_A}^2 \neq {S_B}^2$.

Test statistic

$$F = \frac{n_B S_B^2}{n_B - 1} / \frac{n_A S_A^2}{n_A - 1} = \frac{234.375}{105.263} 2.227$$

Critical value: The tabulated value $F \geq F_{0.05,24,19} = 2.11$.

Decision rule: Hence the critical region is $F > 2.11$.

Conclusion: The observed *F* is in the critical region. We, therefore, reject H_0 and conclude that MLAs from the Congress run state appear more homogeneous in their attitude towards the new economic policy than the MLAs from the Left Front run state.

Let's discuss when to use which test

Case 1: Assumptions
Randomness, normality, known population variances, samples are uncorrelated. **Use z-test**.

Table 4.10 Decision rule for Snedecor's F test for variances proportion tests

Alternative hypothesis	Decision rule
${S^2}_1 > {S^2}_2$	$F > F_{\alpha, N_1-1, N_2-1}$
$S_1^2 < S_2^2$	$F < F_{\alpha, N_1-1, N_2-1}$
$S_1^2 \neq S_2^2$	$F < F_{1-\alpha/2, N_1-1, N_2-1}$

Case 2: Assumptions
Randomness, normality, unknown population variances, samples are uncorrelated. **Use student's t-test**.

Case 3: Assumptions
Randomness, normality, SD of populations unknown but assumed to be equal, uncorrelated small samples. **Use student's t ratio**.

Case 4: Assumptions
Randomness, normality, variances of populations unknown but assumed to be equal, uncorrelated small samples of equal size. **Use f-test**.

Case 5: Assumptions
Randomness, normality, unknown population variances, samples are uncorrelated. **Use student's t-test**.

Case 6: Assumptions
Randomness, normality, variances of populations unknown but assumed to be equal, uncorrelated small samples of equal size. **Use the F-test**.

4.3 ANOVA

Analysis of Variance was first proposed in 1918 by **Sir R. A. Fisher**. But it was formally recognized only after he applied concepts of variance on the crop yield at Broadbalk fields, Rothamsted. In 1925, "**Statistical Methods for Research Workers**" was published in which Fisher formalized the concepts of ANOVA (Fig. 4.12).

Fig. 4.12 Harvesting of Broadbalk Field, [First data source]

Fig. 4.13 Sir Ronald A. Fisher

In hypothesis testing, researchers sometimes would have to analyse how each group of data would respond to several treatments.

For example, a farmer needs to know if he increases the percentage of ammonium in the soil, how would the yield of crops such as maize, jowar and bajra vary?

In most clinical trials, we use ANOVA to test the effect of changes in dosage of medicine on the sugar, blood pressure and thyroid hormone. In both the examples defined above, when there are several groups to compare, it is not possible to compare all possible pairs with t-tests or z-tests. Thus, there is a need for a better test and ANOVA is the solution. ANOVA analyses if there is any significant difference between the means of independent groups across several treatments assigned, simultaneously (Fig. 4.13).

Definition

ANOVA is a procedure in which the difference between several sample means is tested simultaneously. It allows for multiple comparisons while holding the probability of a type I error at a preselected level. ANOVA is also a tool for breaking up the total variance of a large sample or population which consists of many groups.

Caution: Although it is very important to set up the hypothesis before collecting data, many researchers would like to resort to an easier path of considering only two means and applying t-test. For instance, consider a researcher who wants to study four groups of students. Before collecting data, he will not be able to say which two groups are similar/dissimilar and so on.

An important assumption in the analysis of variance is that **all treatments have similar variance**. If there are reasons to doubt this, then the data needs to be transformed before, the test can be done. In practice, there is a simple way to check for homogeneity of variance as follows.

Case Study 1

Consider four streams of an academic programme. A test of mental ability is administered to four random samples, each of size 15 drawn from these streams. The researcher needs to test if there is a significant difference between the means of the four streams.

Suitable tool: **ANOVA**

- **First Reason**: The z- or t-test can be used by considering two means at a time. Thus, there will be 6 pairs for comparison. This involves a lot of time as well as computation. As the sample sizes increase, the number of computations will also go up.
- **Second**, even after all the lengthy computations, there may not be a significant difference between various pairs, which will not help the researcher.
- **Third**, it may turn out that some may be significant at 0.01 level, some at 0.05 level while some may not be significant at any level. This again brings the researcher to a dilemma. Thus, a single composite test is required.
- **Fourth**, the differences between means are not independent of each other. This means, if we know the difference between streams A and B, as well as A and C, we can find the difference between B and C. This shows that one difference is not independent of the other two differences. But both z- and t-tests are used to find differences between pairs of means of two independent random samples. Thus, z- and t-tests are not feasible tools here.

No statistical test can be arbitrarily used and the researcher has to ensure that a specific set of conditions must be met. Such conditions are known as "**Model Assumptions**". For ANOVA, these are

- **Independence** which means the selection of a participant should not be dependent on the selection of any other participant. The most important issue of independence is that **"Observations within or between groups are not paired, dependent, correlated, or associated in any way"** (Glass & Hopkins, 1996, p. 295). Estimates of **type I and type II errors** are not accurate when this condition of independence is violated.
- **Randomness** in selection and assignment is important too, as this provides an equal chance for all units to be selected. All units must be measured only once. For example, if a student has chosen social work and statistics for his study, he should not be measured twice. It would be better to omit such units from the study.
- **Normality**—If we consider a normal sample distribution, the means and variances over repeated samples will be uncorrelated, and therefore, statistically independent. The formulae used to compare groups are based on the mean, median and mode being approximately equal, which is a property of normal distribution.

ANOVA Calculation

See Fig. 4.14.

The variance is divided into two types

1. **Within-group variance**: As the name suggests, this variance explains the variations of elements within each group of data who experience similar experimental conditions.
2. **Between-group variance**: This is the variance that measures the variation across multiple sets of data that occurs due to the interaction of the samples.

Consider a classroom with three different rows of students. All of them are of nearly the same age, educational qualifications and family background. All of them are taught by the same teacher, using the same technique. But still, all of them don't get the same marks. Also, the marks vary in the same group as well. How do we explain this? We call this "**between groups**" that is variance between different rows of students and "**within groups**" meaning variance within each row of students.

Example 14 There were $15 \times 4 = 60$ subjects considered in a study. All the scores of these 60, when added and divided by 4, give us the grand mean. There are 4 groups of 15 people each. The mean of each of these 4 samples is called the **group mean**, and thus, we have 4 group means. These may vary or be almost the same depending on the sample and their scores. The variation of the score of a randomly chosen person of the sample might differ from the group mean—this is **within-group** variance. The variation of the arithmetic mean of a group from the grand mean is known as a **between-group** variance.

The decision criteria for ANOVA

P-value: Refers to the smallest level of significance that would lead to rejection of the null hypothesis. If $\alpha = 0.05$ and the p-value ≤ 0.05, then null hypothesis is rejected. Similarly, if p-value > 0.05, then we fail to reject the null hypothesis.

F-ratio: This is the key factor in the calculation of ANOVA which is the ratio of mean squares between and within the groups. The F-ratio tells us precisely how

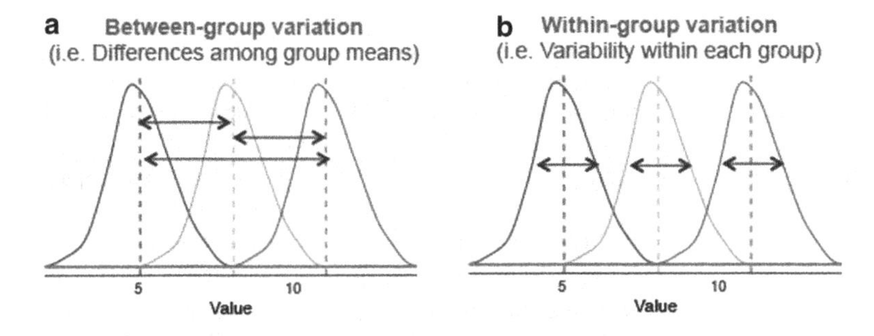

Fig. 4.14 Between and Within sum of squares in a normal distribution [7]

much more of the variation in Y (independent variable) is explained by X (dependent variable). The larger the value of the F-ratio, the lesser the chance of the null hypothesis being right (Fig. 4.15).

Error: While modelling, error or residual refers to the variation between an observed and predicted value. In ANOVA the errors are independent and identically distributed in a normal distribution.

One-Way and Two-way ANOVA

One-way ANOVA examines the effect of variance of a factor on a dependent variable, whereas two-way ANOVA analyses the interdependence of two factors on a dependent variable (Table 4.11).

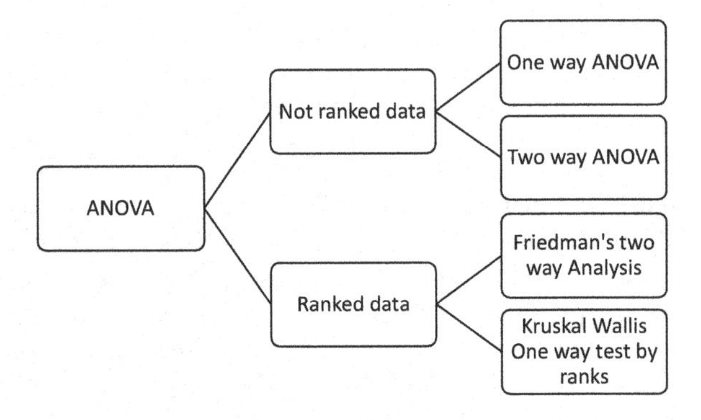

Fig. 4.15 Types of ANOVA based on data types

Table 4.11 One way and two-way ANOVA

	One-way ANOVA	Two-way ANOVA
Definition	A test that allows one to make comparisons between the means of three or more groups of data	A test that allows one to make comparisons between the means of three or more groups of data, where two independent variables are considered
Number of independent variables	One	Two
What is being compared?	The means of three or more groups of an independent variable on a dependent variable	The effect of multiple groups of two independent variables on a dependent variable and on each other
Number of groups of samples	Three or more	There are multiple samples under each variable
Null and alternate hypothesis	H_0: there is no difference between the means and groups	H_0: there is no interaction between the I_1 and I_2 variables
	H_1: there is a difference between the means and groups	H_1: there is interaction between the I_1 and I_2 variables

Math for ANOVA

See Table 4.12.

Sum of Squares Formula

$$\text{SS}_{\text{Total}} = \sum_{j=1}^{p}\sum_{i=1}^{n_j}\left(x_{ij} - \overline{x}\right)^2$$

$$\text{SS}_{\text{between}} = \sum_{j=1}^{p} n_j\left(\overline{x_j} - \overline{x}\right)^2$$

$$\text{SS}_{\text{within}} = \sum_{j=1}^{p}\sum_{i=1}^{n_j}\left(x_{ij} - \overline{x_j}\right)^2$$

Mean Squares Formula

$$\text{MS}_{\text{between}} = \frac{\text{SS}_{\text{between}}}{\text{df}_{\text{between}}}$$

$$\text{MS}_{\text{within}} = \frac{\text{SS}_{\text{within}}}{\text{df}_{\text{within}}}$$

F Formula

$$F = \frac{\text{MS}_{\text{between}}}{\text{MS}_{\text{within}}}$$

The 7 steps to find ANOVA

1. Calculate all the means.
2. Find out the degrees of freedom.
3. Club these values to calculate the sum of squares.
4. Formulate the null and alternative hypotheses.
5. Compute the F statistic.
6. Look at the F statistical table to conclude.

Table 4.12 Math for ANOVA

Source of variation	Sum of squares		Degrees of freedom	Mean squares	F-Statistic
Treatments	SS between	SS_b	$k-1$	$\text{MS}_b = \frac{\text{SS}_b}{k-1}$	$F = \frac{\text{MS}_b}{\text{MS}_w}$
Error	SS within	SS_w	$N-t$	$\text{MS}_w = \frac{\text{SS}_w}{N-k}$	
Total	SS total	SS_T	$N-1$		

Table 4.13 ANOVA table for student's scores

Group	Test scores							
Constant sound	7	4	6	8	6	6	2	9
Random sound	5	5	3	4	4	7	2	2
No sound	2	4	7	1	2	1	5	5

7. Provide a suitable inference for the study based on the conclusion.

In ANOVA, **Post Hoc tests** help us identify which of the several variables are truly independent.

A1. One-Way ANOVA

One-way ANOVA is a simple statistical analysis used to test if there is a statistical difference in the means among two or more independent groups of data. In this test, the data which is divided into two or more becomes the dependent variable and the factors are the independent variables. This type of testing procedure is also called one-factor ANOVA or between-subjects ANOVA.

Case study: A psychologist predicts that students will learn most effectively with a constant background sound as opposed to an unpredictable sound or no sound at all. She randomly divides 24 students into three groups of eight. All students study a passage of text for 30 min. Those in group 1 study with background sound at constant volume. In group 2, students study with noise in the background that changes volume periodically. Those in group 3 study with no sound at all. After studying all the students take a 10-point multiple choice test over the material. Their scores are as follows (Table 4.13).

Suitable tool: One-Way ANOVA.

Reason: The "sound" is the only variable here which may be affecting the scores. For reasons given in case study 1, ANOVA is an appropriate tool.

Solution

H_0 There is no significant difference between tests of various groups.
H_1 There is a significant difference between test scores of various groups.

The ANOVA table is given below:

Sources of variation	Sum of squares	Degrees of freedom	Mean sum of squares	F
Within groups	87.88	21	4.18	3.59
Between groups	30.08	2	15.04	

Table value of F at (2, 21) at 0.05 = **3.4668.**

Table 4.14 The data that shows the bacteria count on shipments 1-5 and the samples 1-6

Sources of variation	Sum of squares	Degrees of freedom	Mean sum of squares	F
Within groups	803	4	200.75	9.01
Between groups	557.17	25	22.287	

Conclusion: Since the F value > table value of F, the researcher has enough reasons to reject the null hypothesis. There is a significant difference in the test scores which is definitely not due to chance.

Conclusion: The psychologist can conclude that hypothesis H_1 may be supported. The means he/she predicted and the constant music group has the highest score.

However, the significant F only indicates that at least two means are significantly different from one another, but the researcher will not know the means of which specific pairs are significant from each other. This is understood only through a post hoc analysis like **Tukey's HSD**.

Case Study

A study was reported by Hogg and Ledolter (1987), of bacteria counts in shipments of milk. There were five shipments and for each shipment, bacteria counts were made from 6 randomly selected cartons of milk. The question to be answered is whether some shipments have higher bacteria counts than others. The data shows the bacteria count on shipments 1–5 and samples 1–6 (Table 4.14).

Suitable tool: One-way ANOVA.

Reason: The bacterial count is the only variable here that may be affecting the cartons of milk. For reasons given in case study 1, ANOVA is the suitable choice.

Solution

H_0 There is no significant difference between milk cartons with respect to bacterial count.

H_1 There is a significant difference between milk cartons with respect to bacterial count.

The ANOVA table is given below (Table 4.15).
Table value of F at (4, 25) at 0.05 = **2.7587**.

Conclusion: Since the F value > table value of F, the researcher has enough reasons to reject the null hypothesis. There is a significant difference in the bacterial counts which is definitely not due to chance.

What next? In an analysis of variance, when the observed value of F exceeds the expected theoretical value, it means

- The samples are from different populations.

- It also tells us that there are differences between treatments in the experiment as a whole but it does not tell us which treatments differ from one another. To check which pair has different means, we apply a student's t-test assuming that all populations share a common variance. The critical difference in method is then used to find the pairs.

Case Study

How does an MBA major affect the number of job offers received? An MBA student randomly sampled four recent graduates, one each in finance, marketing, and management, and asked them to report the number of job offers. Can we conclude at the 5% significance level that there are differences in the number of job offers between the three MBA majors? (Table 4.16).

Suitable tool: One-way ANOVA.

Reason: The "MBA major" is the only variable here that may be affecting the number of jobs received.

Solution

The one-way ANOVA model is given by

$$y_{ij} = \mu + \alpha_i + \epsilon_{ij} \; i = 1, 2 \ldots k \text{ and } j = 1, 2, \ldots n_i$$

y_{ij} Number of jobs offers recorded by the jth unit of the ith major
α_i Effect of the $i$thmajor
ϵ_{ij} Random error.

Table 4.15 ANOVA table for samples in each shipment

		Shipment				
		1	2	3	4	5
Samples	1	24	14	11	7	19
	2	15	7	9	7	24
	5	21	12	7	4	19
	4	27	17	13	7	15
	5	33	14	12	12	10
	6	23	16	18	18	20

Table 4.16 Table showing job offers for each department

Finance	Marketing	Management
3	1	8
1	5	5
4	3	4
1	4	6

Table 4.17 ANOVA table for jobs in each department

ANOVA						
Source	SS	df	MS	F	*P*-value	*F* crit
Between groups	26	2	13	**4.824**	0.0376	**4.256**
Within groups	24.5	9	2.69			
Total	50.25	11				

Null hypothesis

H_0^M There is no difference in the average number of job offers between the 3 MBA majors. $\alpha_i = 0$ for all values of $i = 1, 2, \ldots k$.

Alternative hypothesis

H_1^M There is a difference in the average number of job offers between the 3 MBA majors. $\alpha_i = 0$ for atleast one i value.

Test statistic

$$F = \frac{\text{MMSS}}{\text{MESS}} \sim F_{k-1,N-k}$$

Level of significance: $\alpha = 0.05$.
Decision rule: Reject H_0^M if $F \geq F_{\alpha,k-1,N-k}$

Calculation

See Table 4.17.

Inference: Since 4.8247 > 4.256495, we reject H_0^M

Conclusion: There is a difference in the average number of jobs offers between the 3 MBA majors.

A2. Two-Way ANOVA

In one-way ANOVA, there was only one experimental variable. But sometimes, we come across situations that require a simultaneous study of two experimental variables. In such scenarios, two-way ANOVA is ideal to use. When there are more than two experimental variables studies simultaneously, we use MANOVA. Analysis of covariance is called ANCOVA.

Example 15 Consider a class that is taught by a teacher who follows different methods of teaching, such as

- A lecture.
- Demonstrations.
- Collaborative learning and

- Case studies.

In this case, the method of teaching is the experimental/independent variable which has to be applied at four levels. Consider four groups of students selected randomly from the class. These four groups are taught by the same teacher but using different methods. A test is given by the teacher, and the mean score of these groups is computed. If we are interested in knowing the significance of differences between the means of these groups, we use one-way ANOVA as there is only one experimental variable, namely method of teaching (Fig. 4.16).

Now, consider another independent/experimental variable such as three types of schools in which this experiment is conducted. These school systems can be State syllabus, ICSE and CBSE. The experiment now has 4 × 3 groups. That is four groups in each type of school. This is a case where two-way ANOVA is used.

Here, the population variance is divided into three types:

(i) Variance due to methods of teaching.
(ii) Variance due to the syllabus.
(iii) Interaction variance or the residual variance which can exist due to many factors like chance, lack of training in teachers, etc.

Resolving confusion between contingency tables and two-way ANOVA:

Two-way ANOVA is not the same as a contingency table. In both, we classify subjects according to their values for two categorical variables. In two-way ANOVA, the value of some quantitative response variable is recorded, whereas in the contingency table, we just count the number of variables in each cell and record it in that particular cell. The following table will illustrate it better (Table 4.18).

Assumptions for two-way ANOVA are the same as that of one way with an additional requirement. The number of observations should be the same for all groups.

Case Study 1

A research study was conducted to examine the impact of eating a high protein breakfast on adolescents' performance during a physical education physical fitness

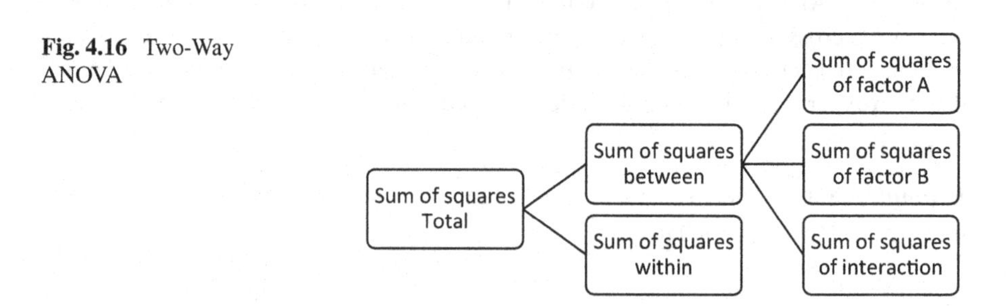

Fig. 4.16 Two-Way ANOVA

Table 4.18 Variable types and method to be used

Explanatory variable	Response variable	Method to be used
Categorical variable	Categorical variable	Contingency table
Categorical variable	Quantitative variable	ANOVA
Quantitative variable	Quantitative variable	Regression

Table 4.19 Protein levels in males and females

Group	High protein	Low protein
Males	10	5
	7	4
	9	7
	6	4
	8	5
Females	5	3
	4	4
	6	5
	3	1
	2	2

test. Half of the subjects received a high protein breakfast, and half were given a low protein breakfast. All of the adolescents, both male and female, were given a fitness test with high scores representing better performance. Test scores are recorded below (Table 4.19).

Suitable tool: **Two-way ANOVA**

Reason: Gender, as well as protein level, is considered in this research question.

Solution

The null hypotheses are as follows:

Null Hypothesis: The population means of the protein level are equal.
(This is like the one-way ANOVA for the column factor.)

Alternate hypothesis: The population means of the gender are not equal.
(This is like the one-way ANOVA for the row factor.)

There is no interaction between the two factors. (This is similar to performing a test for independence with contingency tables.)

A two-way ANOVA table is given below (Table 4.20).

Table 4.20 ANOVA table for protein levels in males and females

Source of variation	Sum of squares	Degrees of freedom	Mean sum of squares	F
Factor 1: protein level	20	1	20	8.89
Factor 2: gender	45	1	45	20
Interaction between factor 1 and factor 2	5	1	5	2.22
Within	36	16	2.25	

Conclusion: There seem to be significant main effects for both protein level and gender. There was not a significant interaction effect Based on this data, it appears that a high protein diet results in a better fitness test score.

Case Study 2

A study examining differences in life satisfaction between young adult, middle adult and older adult men and women was conducted. Each individual who participated in the study completed a life satisfaction questionnaire. A high score on the test indicates a higher level of life satisfaction. Test scores are recorded below (Tables 4.21 and 4.22).

Conclusion: There are significant main effects for age and gender. There is no interaction effect. It appears from the data that older adults have the highest life satisfaction and younger adults have the lowest life satisfaction.

A3. Kruskal–Wallis One-Way Analysis of Variance by Ranks

This is a very powerful test similar to a one-way analysis of variance used for ranked data. For measurement by ranks of k independent samples. The null hypothesis, H_0 is tested, if **k** samples come from the same population. That is, the average rank is the same for all the k samples.

Table 4.21 Life satisfaction levels among people

Group	Young adult	Middle adult	Older adult
Males	4	7	10
	2	5	7
	3	7	9
	4	5	8
	2	6	11
Females	7	8	10
	4	10	9
	3	7	12
	6	7	11
	5	8	13

Table 4.22 ANOVA table for life satisfaction levels among people

Sources of variation	Sum of squares	Degrees of freedom	Mean sum of squares	F
Factor 1: age	180	2	90	49.09
Factor 2: gender	30	1	30	16.36
Interaction between factor 1 and factor 2	0	2	0	0
Within	44	24	1.83	

The scores of **k** samples are first considered as belonging to the same series. The rank of each score is then ascertained from the k combined sample series. The smallest score is assigned rank 1 and the largest to **n**, where $n = n_1 + n_2 + \cdots + n_{k.}$ For each of the samples, the sum of ranks is worked out.

If H_0 is true, that is, k samples are drawn from the same population, then Kruskal–Wallis statistic, designated as **H,** follows a distribution similar to chi-square with degrees of freedom, df $= k - 1$ provided that the size of all k samples is not too small. **H** is defined as

$$H = \frac{12}{N(N+1)} \sum_{j=1}^{k} \frac{R_j^2}{n_j} - 3(\mathrm{N}+1), \text{ where } N = \sum n_j,$$

R_j = sum total of ranks in the jth sample (column). Table of chi-square may be used for $n_j > 5$ for ascertaining the probability of obtaining a value of λ^2 from the table for a level of significance α. For $k = 3$ and $n = 5$, the sampling distribution of H is not close to that of λ^2.

Case Study
Three cricket teams completed four 20-over matches. The following table shows how the teams fared in the 4 innings they played. Does the three groups differ in their average performances? (Table 4.23).

Suitable Tool: Kruskal–Wallis one-way ANOVA.

Reason: The data requires a test for tackling simultaneously 3 independent samples. Cricket scores can be easily be expressed in ordinal measurements. Kruskal–Wallis one-way ANOVA seems appropriate for testing the differences.

Solution

1. **Null hypothesis $\mathbf{H_0}$**: There is no difference among the teams in their average performances.
   Alternate hypothesis $\mathbf{H_1}$: The teams differ in their average performances.
2. For $k = 3$ and sample size not exceeding 5, we select 0.05 as the level of rejection of H_0.
3. **Calculations**

Table 4.23 Table showing scores of 3 cricket teams in 4 innings

A	B	C
150(7)	140(5)	170(11)
122(2)	120(1)	170(11)
130(3)	132(4)	145(6)
160(9)	170(11)	145(8)

i. Assign ranks to every individual score assuming all of them belong to the same series. Rank 1 is given to the lowest value, rank 2 to the next smallest value and so on. For the tied observations, the average rank is given. The ranks are written in parentheses against the observed scores. (In this case, three values are assigned rank 11 as the arithmetic mean of ranks 10, 11 and 12 is 11)
ii. Find the sum of ranks separately for each sample. Thus, $\sum \text{RA} = 21$, $\sum \text{RB} = 21$, $\sum \text{RC} = 36$.
iii. Correction for ties: Tied observations may influence H. Effect of ties is corrected by dividing H by the correction factor.

where t = number of tied observations in a tied group of scores, n = number of observations of k samples taken together and $T = t3 - t = 33–3 = 24$. The correction factor in the example is

$$1 - [24/(123 - 12)] = 0.986.$$

$$H = \frac{12}{N(N+1)} \sum_{j=1}^{k} \frac{R_j^2}{n_j} - 3(N+1)$$

$$= \frac{12}{12 * 13} * \frac{21^2 + 21^2 + 36^2}{4} - 3 * 13 = 3$$

$$\text{Corrected } H = 3.0/0.986 = 3.0429$$

From the table, we see that $H_{0.5} = 5.6923$ has a probability of 0.049.
The observed $H = 3.042$ is much smaller than the tabulated value of $H_{0.5}$.

Conclusion: So, H is outside the critical region and we accept H_0 and conclude that the average performances of the three teams are not different.

A3. Friedman Two-Way Analysis of Variance by Ranks, Related Samples

When data from k-related samples is at least in an ordinal scale, two-way analysis of variance by ranks, suggested by Friedman, is very useful for testing the null hypothesis that k-related samples come from the same population.

The samples may be matched either by studying the same group under different treatment conditions (LSD) or several individuals may be matched with respect to some major variables other than the one being experimented upon (block design), which may influence measurements. The samples so selected are randomly assigned to different treatment groups. Therefore, K-related samples are equal in size.

As in RBD, we have r-rows and k-columns. The individuals in a particular row are matched before they are offered any treatment. In this way, they constitute blocks. The basic assumption is that since each row consists of individuals who are matched, the scores of these subjects may be considered as scores of a single subject but under different treatment conditions.

The scores in each row are ranked separately. In absence of treatments, the scores within a block should have been identical and therefore their ranks equal. The ranks in each row range from a low of 1 to a high of k, k being the number of columns. This is similar to saying that had there been no differential treatment effect on the dependent variable under measurement, rank distribution in a particular column should be the same as the distribution of ranks in the same population. **Friedman two-way analysis of variance is intended to test if the rank totals of different columns (samples) are the same**. If the null hypothesis is true, the distribution of ranks in each column would be a matter of chance and the ranks 1, 2 ... k are likely to appear with about equal frequency in all the columns. The test Friedman uses are denoted as λ_r^2 and are approximately distributed as chi-square with df $= k - 1$.

$$\lambda_r^2 = \frac{12}{nk(k+1)} \sum_{j=1} R_j^2 - 3n(k+1)$$

where n = number of rows, k = number of columns and R_j = sum of ranks in the jth column. When the number of rows and columns is not too small, λ_r^2 may be compared with the tabulated value of λ_r^2 for $k = 3$ and $n = 2$–9 and for $k = 4$ and $n = 2$–4 are given in Table.

Case Study

A group of 5 students participated in a cancellation test under three different lighting conditions—25, 60 and 150 W. They were asked to cancel every "*i*" followed or preceded by "*t*" from a standard passage in a fixed time. Three comparable passages were used for three sessions of testing. Does the intensity of illumination has any significant effect on the test scores? The number of errors committed is the score of an individual (Table 4.24).

Suitable Tool: Friedman's two-way analysis of variance.

Reason: Measurement of error in the experiment does not have the strength of a continuous scale. However, the error scores can be expressed quite easily on an ordinal scale. Since the samples are matched (each student appears in all three sessions), we choose Friedman's two-way analysis of variance in preference to the parametric F-test.

Table 4.24 Ranks for cancellation test

Student	25 W	60 W	150 W
1	17	8	13
2	14	10	10
3	10	12	15
4	20	15	18
5	10	13	8

Solution

1. H_0: Variation in intensity of illumination within the range specified has no effect on an individual's performance.
2. H_1: Variation in intensity of illumination has an effect on an individual's performance.
3. For $k = 3$ and $n = 5$, we are in a position to find out the exact probability of obtaining λ_r^2 with df $= k - 1$ with reference to the table.

   Rejection Region: It consists of all values of λ_r^2, which are so large that the probability associated with its occurrence under H_0 is equal to or less than 0.05.
4. **Calculations**: Scores within each row are ranked separately and the sum of the ranks from each column is obtained (Table 4.25).

For tied ranks, no correction is needed for λ_r^2.

$$\lambda_r^2 = \frac{12}{nk(k+1)} \sum_{j=1} R_j^2 - 3n(k+1)$$

$$= \frac{12}{5 * 3 * 4}(144 + 75 + 90.25) - (3 * 5 * 4) = 1.3$$

5. **Decision**: With reference to the computations, we find that $\lambda_r^2 = 1.3$ has a probability greater than 0.522. So, we accept H_0 and conclude that the intensity of illumination has not affected the performance of the students significantly.

4.4 Post Hoc Tests

These are the tests that are performed after ANOVA to identify the different groups that have statistically significant differences in their means. Welch's ANOVA is considered as an **Omnibus Post Hoc test for ANOVA**. The most commonly used Post Hoc test is **Tukey's method.**

There are two different testing methods are:

i. Adjusted p-values.

Table 4.25 Cancellation test of 5 students over different lighting condition

Student	25 W	60 W	150 W
1	3	1	2
2	3	1.5	1.5
3	1	2	3
4	3	1	2
5	2	3	1
Total	**12**	**8.5**	**9.5**

Tukey Simultaneous Tests for Differences of Means

Difference of Levels	Difference of Means	SE of Difference	95% CI	T-Value	Adjusted P-Value
B - A	-6.17	2.28	(-12.55, 0.22)	-2.70	0.061
C - A	-1.75	2.28	(-8.14, 4.64)	-0.77	0.868
D - A	3.33	2.28	(-3.05, 9.72)	1.46	0.478
C - B	4.42	2.28	(-1.97, 10.80)	1.94	0.245
D - B	9.50	2.28	(3.11, 15.89)	4.17	0.002
D - C	5.08	2.28	(-1.30, 11.47)	2.23	0.150

Individual confidence level = 98.89%

Fig. 4.17 Tukey post hoc test [8]

ii. Simultaneous confidence intervals.

Example 16 There are 4 materials A, B, C and D for which we need to determine their strength. We performed a one-way ANOVA test, and the *p*-value obtained was 0.04. This undoubtedly rejected the null hypothesis that all 4 means were equal. But now, we do not know which combination of materials have significant differences in their means. Therefore, Tukey Post Hoc tests were done, and this is how we interpret the results in both the methods discussed above.

The result of Tukey's test is shown in Fig. 4.17. Consider 4 groups A, B, C and D. In the **"Adjusted p-value"** column, identify the value that is lesser than the *p*-value. The corresponding groups to this value are said to have means that are statistically significant from each other. We generally use a *p*-value of 0.05; therefore, the groups "D and B" have significantly different means. It is also evident because the difference in their means is 9.5 which is the maximum in the column **"Difference of Means"**.

Simultaneous class intervals

For the same example, we have plotted the class intervals in 95% class intervals which are shown in the fourth column in Fig. 4.18. Here the zero value indicates that the group means are equal. The groups that do not include zero in the interval are said to be statistically significant from the others. By this, we can once again confirm that the means of groups D and B differ from each other.

4.5 Non-parametric Tests

Definition

Nonparametric tests are also called distribution-free tests with no underlying assumptions too. This is because they can be used with those data values which do not conform to certain assumptions. They can be used for very small samples as well as on those measured on a ratio or interval scale.

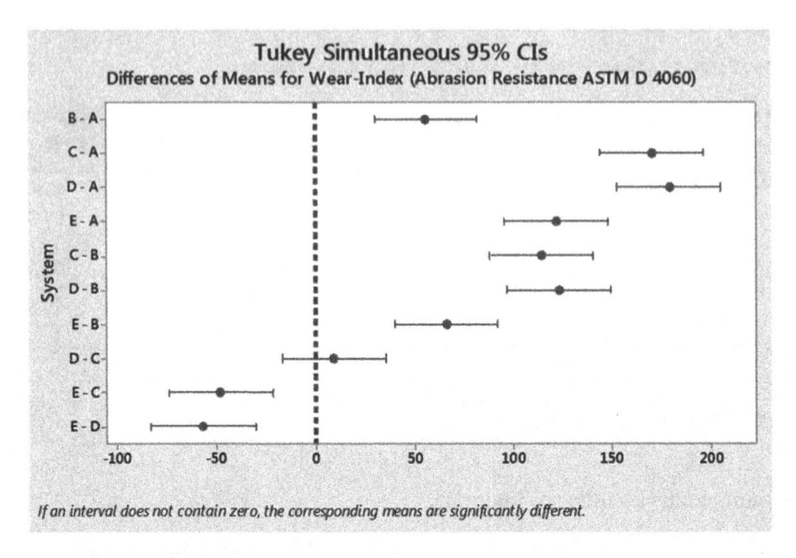

Fig. 4.18 Tukey's method of simultaneous 95% CI class intervals [9]

A recent series of papers by **Charles T. Perretti** and collaborators have shown that non-parametric forecasting methods can outperform parametric methods in noisy nonlinear systems. Such a situation can arise because of two main reasons.

i. The instability of parametric inference procedures in chaotic systems which can lead to biased parameter estimates.
ii. The discrepancy between the real system dynamics and the modelled one, a problem that Perretti and collaborators call "**The True Model Myth**" (Jabot 2015).

Seigal (1956) gave the following conditions for using parametric tests like *Z*, *t* and *F*:

1. The population from which the samples have been drawn should be normally distributed. (Assumptions of normality).
2. The variables involved must be measured in an interval or ratio scale.
3. The observations must be independent. The exclusion or inclusion of any case in the sample should not unduly affect the results of the study.
4. These populations must have known variance, or, in special cases, must have a known ratio of variance. This is called homoscedasticity.

Caution: As compared to parametric tests, nonparametric tests are easier to use. But their usage has to be restricted to such situations where parametric tests conditions listed above are not met.

Conditions for using non-parametric tests

1. The size of the sample is very small.
2. The normality of the distribution of scores in the study is not certain.

3. When data is measured in either ordinal or nominal scales.

Some commonly used non-parametric tests for nominal data

See Table 4.26.

Case I: One Sample, Two Category Observations
Here, again, two cases are possible:

I. A population consists of only two classes: married-single, Hindu-, non-Hindu, literate-illiterate, public sector and private sector, etc. We may also have observations consisting of only two categories: pass-fail, agree-disagree, yes–no, voted-abstained, etc.

That is, the observations fall on either of two discrete categories such that if p is the observed proportion in one category, $(1 - p) = q$ is the proportion in the other.

Check

A. Is the sample small? **Use binomial or chi-square test**.

Table 4.26 Commonly used non-parametric tests for nominal data

Research issue	Parametric test	Non-parametric test
To compare mean value for some variable of interest in two samples	*t*-test	1. Wald-Wolfowitz runs test
		2. Mann–Whitney U test
		3. Kologorov–Smirnov two sample tests
Differences between independent multiple groups	ANOVA/MANOVA	1. Kruskal–Wallis
		2. Median test
To compare two variables measured in the same sample	ANOVA/MANOVA	1. Kruskal–Wallis
		2. Median test
To compare variables if more than two variables are measured in same sample	Repeated measures ANOVA	1. Friedman's two-way ANOVA
		2. Cochran's Q
To study the relationship between two categorical variables	Correlation coefficient	1. Spearman's ρ
		2. Kendall's τ
		3. Coefficient Gamma
		4. Chi-square
		5. Phi coefficient
		6. Fisher's exact test
		7. Kendall's coefficient of concordance
To compare two variables if more than two variables are measured in the same sample	Repeated measures ANOVA	1. Friedman's two-way ANOVA
		2. Cochran's Q

Fig. 4.19 Binomial tests [10]

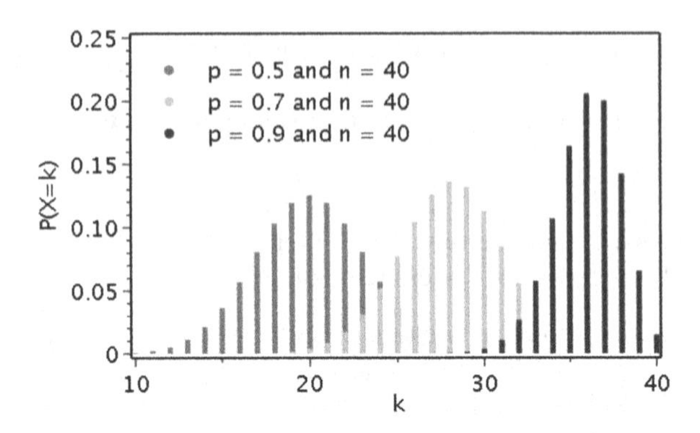

B. Is the sample large? (When the sample is large, the binomial approaches Poisson distribution.) **Use the z-test.**

II. The populations, more specifically the responses of individuals in the population, fall into two or more categories (Fig. 4.19).

The binomial test

When considering type I situations, the binomial test may be considered appropriate though the chi-square test can also be used. The binomial distribution, we know, is a sampling distribution of two mutually independent discrete events, head and tail, literate and illiterate, etc. The binomial distribution gives certain values that are expected to occur under null hypothesis H_0. It tells us if the observed proportion of occurrence of a particular event was in tune with what was expected of a population. That is, the observed value of the proportion **p** is different from **P** in the population under the alternative hypothesis.

The probability of obtaining x number of objects in one category and $(n - x)$ number of objects in another category out of a total of **n** objects is given by

$$P(X) = \frac{n!}{x! * (n - x)!} p^x * q^{n-x} (0 \; to \; n)$$

where p is the proportion of observations in the specified category.

Large Sample

For large samples, binomial distribution approaches the normal distribution. The tendency is more conspicuous and rapid as **p** approaches **0.5** than when it is close to 0 or 1. When **p** lies in either of the extremities, **n** must be very large before binomial distribution can be represented by the normal distribution.

There is no clear direction as to the size of the sample that must be achieved for using normal approximation when p is in either of the extremities. Generally, however, it is recommended that the value of **npq** must be at least 9 before a statistical test based on normal approximation is applied.

For example, if p is 0.05, the sample size should be at least as large as 190 ($n \times 0.05 \times 0.95 = 9$). When these conditions are met, the sampling distribution of x is approximately normal with mean $= np$ and SD $= \sqrt{npq}$ on H_0 may be tested by

$$Z = \frac{x - \mu}{\sigma_x} = \frac{x - nP}{\sqrt{nPQ}}$$

A binomial distribution is a discrete distribution, whereas a normal distribution is for a continuous variable. For a cumulative distribution for the said interval, the cumulative frequency is for the upper end of the interval. Z then becomes $Z = \frac{(x \pm 0.5) - nP}{\sqrt{nPQ}}$.

Case Study

Opinions of 20 headmasters were sought if English should be reintroduced at the primary level. Sixteen opined in favour. Does the data suggests that reintroduction of English at the primary level has a strong ground?

Suitable Tool: Binomial test-large samples.

Reason: The data is in discrete categories, and a single sample has been employed and since p and q are not too extreme and the size of the sample is not too small, we may use a normal approximation of the binomial with correction for continuity. The test selected is z.

Solution

Null hypothesis $\mathbf{H_o}$: The proportion of headmasters favouring reintroduction of English and the proportion against it do not differ significantly from $\mathbf{p = q = 0.5}$.

Alternative hypothesis $\mathbf{H_A}$: The proportion of headmasters favouring reintroduction of English and the proportion against it differ significantly. The outcome equals some claimed value or $p > q$.

Let $p = 16/20 = 0.8$ represent the proportion of those favouring reintroduction. $q = 4/20 = 0.2$ the proportion opposing it. $n = 20$,

Our x could either be 16 or 4, and both will yield the same value for a one-tailed test.

$$Z = \frac{(x \pm 0.5) - nP}{\sqrt{nPQ}} = \frac{(16 - 0.5) - 20 * 0.5}{\sqrt{20 * 0.5 * 0.5}} = 2.4597$$

or

$$Z = \frac{(4 + 0.5) - 20 * 0.5}{\sqrt{20 * 0.5 * 0.5}} = 2.4597$$

Chi-squared Test

The chi-squared test is used when there is one sample and more than two categories of variables are to be tested. The shape of a chi-square test looks like the one in

Fig. 4.20. Like a positively skewed normal distribution. In certain situations where observations are obtained on several discrete categories; for example, students may be asked to indicate their preferences for games listed. A number of families in a village by their caste categories, etc. All these examples show that the observations are on a nominal scale and the researcher's objective is to study if the frequencies associated with these categories are significantly different.

Definition of chi-square

If X is normally distributed with mean μ and variance $\sigma^2 > 0$, then:

$V = (X - \mu/\sigma)^2 = Z^2$ is distributed as a chi-square random variable with 1 degree of freedom.

Example 17 Let one variable be the gender of the voter—male, female and the other is the voting preference—Party A, B, C, etc.

Chi-square test for independence can be used to determine whether gender is related to voting preference. The following conditions are to be met for applying chi-square test.

- The sampling method is simple random sampling.
- Each population is at least 10 times as large as its respective sample.
- The variables under study are categorical.
- If sample data is displayed in a contingency table, the expected frequency count for each cell of the table is at least 5.

Test for Goodness of Fit and Test of Independence

In a normal distribution scenario, sometimes we might come across situations where we need to check if the sample data we have collected fits an expected data set from a population. In such situations, chi-square test for goodness of fit is used.

This chi-square test is not appropriate for continuous data. This test checks for the differences in the observed and expected values. Most machine learning models

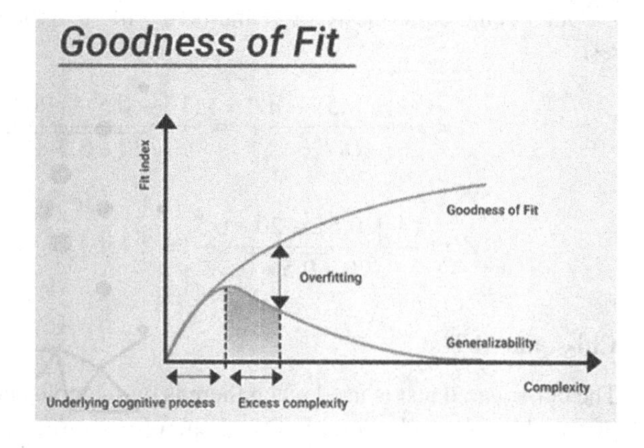

Fig. 4.20 Goodness of fit [11]

use this test to confirm if the trained data set is a good fit for the expected data/test data set. The machine learning algorithm learns and relearns constantly until the trained and test data sets fit appropriately. An analyst formulates a hypothesis about the numbers expected for each of the response categories.

In social sciences, hypotheses are framed from personal experience, intuition and insight. Consistencies in data like birth/ death rates, growth of production, etc., are used for developing hypotheses. Even experimental data can be the basis for hypothesis testing.

Chi-square statistic, denoted by χ^2, is a very important statistic which is concerned with the distribution as a whole and not concerned with the values of the parameter, unlike z-test or t-test. Also, it is not confined to the study of parametric distributions like binomial, Poisson etc.

Karl Pearson proved that $\boldsymbol{\chi^2} = \frac{\sum_{i=1}^{k}(O_i - E_i)^2}{E_i}$ follows chi-square distribution for large values of **n**.

O_i—denotes the observed values, and E_i denotes the expected values. Though this distribution can be applied to both continuous and discrete data, χ^2 is a continuous distribution that is unimodal and its range consists of the positive half of the real axis.

Decision Criteria

Use of chi-square as a test of "goodness of fit"

We can find expected frequencies from some theoretical concepts. Then calculate the goodness of fit of the observed values to expected values.

Steps

1. Set up the null and alternative hypotheses.
   - **Null hypothesis**: There is no significant difference between the expected and observed frequencies. OR The model is a good fit.
   - **Alternative hypothesis**: There is a significant difference between the observed and expected values. OR The model is not a good fit.
2. Computation of expected frequencies (E) and value of $\chi 2$. The test statistic is given by:

$$\chi^2 = \sum \frac{(f_o - f_e)^2}{f_e}$$

3. **Degrees of freedom** df $= (r - 1)(c - 1)$
   where **r** denotes the number of rows in the contingency table and **c** denotes the number of columns in the contingency table.
4. **Decision Criteria**: The table value of $\chi 2$ is read from the table for a predetermined significance level. The $\chi 2$ value obtained is then compared against the tabulated value χ_α^2 for $k - 1$ degrees of freedom.

5. If the calculated value of $\chi 2$ is greater than the table value, it is taken as significant, and consequently, the null hypothesis is rejected. This means that the difference between the observed and expected frequencies is significant and cannot be explained as a matter of chance or by sampling fluctuation.

In case the computed value of $\chi 2$ is less than the critical value of $\chi 2$, then it is called non-significant, and consequently, the null hypothesis is not rejected. That is, we may agree that the difference between observed and expected frequencies is not significant. It may occur by chance or due to sampling fluctuation. When a null hypothesis is not rejected, the theoretical distribution based on some hypothesis (equal probability or normal distribution) may be considered to be a good fit or substitute for the distribution based on experimentally obtained frequencies.

Case Study 1

A public opinion poll surveyed a simple random sample of 1000 voters. They were chosen from a huge population from the voter database. Respondents were classified by gender (male or female) and by voting preference (local party, regional party, national party). Results are shown in the contingency table below (Table 4.27).

Suppose the research objective is to test for gender bias, more specifically, does the voting preferences of men differ significantly from that of women?

Why is chi-square appropriate in this case?

- The sampling method was simple random sampling.
- Each population was more than 10 times larger than its respective sample variables under study are categorical.
- The expected frequency count was at least 5 in each cell of the contingency table.

Case Study 2

A drug manufacturing company conducted a survey of customers. The research question is: Is there a significant relationship between packaging preference (size of the bottle purchased) and economic status? There were four packaging sizes: small, medium, large and jumbo. Economic status was: lower, middle, and upper. The following data was collected (Table 4.28).

Why is chi-square appropriate in this case?

The sampling method was simple random sampling. Each population was more than 10 times larger than its respective sample variables under study are categorical. The expected frequency count was at least 5 in each cell of the contingency table.

Note

Table 4.27 Voting preferences among males and females

Gender	Voting preference		
	Local party	Regional party	National party
Male	200	150	50
Female	250	300	50

Table 4.28 Packaging preferences and economic status

	Lower	Middle	Upper
Small	24	22	18
Medium	23	28	19
Large	18	27	29
Jumbo	16	21	33

1. Chi-square analyses are only valid when the actual observations within the cells are independent, this is not the same as testing whether the variables are independent or not (i.e. the purpose of the test). Observations are not independent when the grand total is larger than the number of subjects, that is when a variable occurs in more than one category.

Consider the following example:

Example 18 A student categorized the level of activities of five animals on four days and obtained the following results (Table 4.29).

Though the number of animals is only 5, there are a total of 20. That is because; each animal is contributing 4 values to the total. The activities exhibited by animals are not independent, as an animal showing low activity on a particular day has a good chance of showing the same on other days too. So, checking for N is one way to check for independence.

2. Another assumption is that the expected values are normally distributed; this assumption breaks down when the expected values are small Thus, one should be cautious using the chi-square test when the expected values are small; thus, caution should be exercised when expected values are less than 5.
3. All outcomes (occurrences and non-occurrences) should be considered in the contingency table, and no values must be left out.

Errors/confusions which can occur:

Example 19 Students are asked to answer 5 out of 10 questions. Does the response pattern of 24 students listed below suggests any pattern of preference? (Table 4.30).

Suitable Tool: Chi-square test.

Table 4.29 Activities of 5 animals on four days

High	Medium	Low	Total
10	7	3	20

Table 4.30 Students and their preference of patterns

Questions	Number of students
1	15
2	7
3	22
4	20
5	10
6	7
7	6
8	12
9	16
10	5

Reason: The data in multiple discrete categories and from a single sample requires χ^2 analysis since the hypothesis under scrutiny demands a comparison of observed and expected frequencies.

Solution

Null hypothesis: $\mathbf{H_0}$: No particular preference pattern exists. The proportion of answering to any question Q_i is the same as the proportion of answering to another question Q_j symbolically,

$$P_i = P_j$$

Alternative hypothesis: $\mathbf{H_1}$ Some questions are given preference over others.

$$p_i \neq P_j$$

Calculations: $X^2 = \sum \frac{(f_0 - f_e)^2}{f_e}$.

where f_0 is the observed frequency and f_e is the expected frequency. The X^2 value obtained is then compared against the tabulated value X_α^2 for $k - 1$ degrees of freedom. Since there are 120 responses, $n = 120$.

Each student is supposed to answer 5 questions out of the 10 given; hence under the null hypothesis, each question will have an equal preference, and since there are 24 students, we expect exactly 12 students to answer any particular question. Therefore, f_e for all the questions from 1 to 10 is 12.

The computed chi-square is

$$X^2 = \frac{1}{12}\left[(5-12)^2 + (7-12)^2 + \cdots + (16-12)^2 + (5-12)^2\right] = \mathbf{27.33}$$

Degrees of freedom: df = $k - 1 = 10–1= \mathbf{9}$.

Rejection region: All values greater than or equal to the tabulated value $X^2_{0.01,9} = 21.67$.

The calculated value of X^2 exceeds the tabulated value; hence, we reject the null hypothesis in favour of the alternative hypothesis.

Conclusion: The students have more frequently answered some questions than others.

Case II: When there are two independent samples.

First check,

- Is the sample small? Use **Fisher's exact probability test.** (Explained next) which can be used for ordinal data too.
- Is the sample large? Use the **chi-square test**, which can be used for nominal, ordinal and interval scale data.

Fisher's Exact Probability Test

When two independently drawn samples with two categories each are small, Fisher's exact probability test is an extremely useful non-parametric test. This can be used for both, ordinal and nominal discrete data. The squares below represent different sizes of data. Fisher's exact probability test measures if there is a significant difference in the observed data size than what is expected. This test basically checks for non-randomness in categorical variables. Continuous variables can also be made discrete when they are classified as below and above a particular value. Every element of the two groups can be placed in mutually exclusive classes. This gives rise to the following 2 × 2 contingency table (Fig. 4.21; Table 4.31).

The objective is to see if the two groups differ in proportion based on the categories in which they belong. The exact probability of observing a particular set of frequencies in a 2 × 2 table is given by

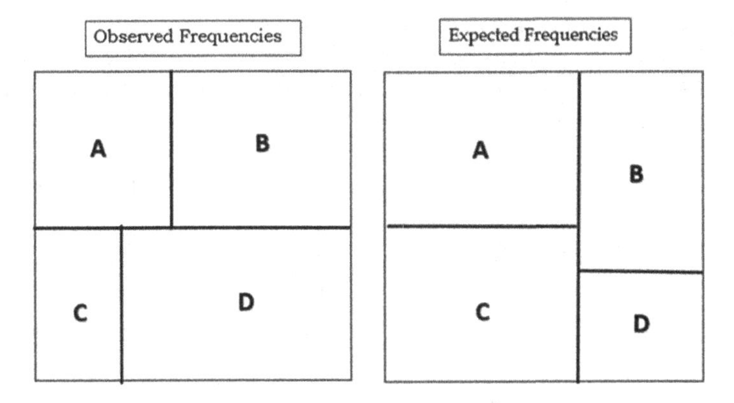

Fig. 4.21 Fisher's exact tests

Table 4.31 Tabular representation of Fisher's exact test

	Type 1	Type II	Total
Sample 1	A	B	$A + B$
Sample 2	C	D	$C + D$
Total	$A + C$	$B + D$	N

$$P = \frac{(A+B)! + (A+C)! + (B+D)! + (C+D)!}{n!A!B!C!D!}$$

Example 20 Compute **p**, given the following results of clinical research on two treatment types conducted on 69 patients (Table 4.32).

Null hypothesis: There is no difference between the two treatments.

Alternative hypothesis: There is no difference between the two treatments.

Test Statistic

$$P = \frac{(A+B)! + (A+C)! + (B+D)! + (C+D)!}{n!A!B!C!D!} = 0.00605$$

This means that the exact probability that these 69 observations should fall in the four cells in the manner they did, which is **0.00605**.

Conclusion: A p-value less than or equal to 0.05 means that our result is statistically significant, and we can trust that the difference is not due to chance alone.

Two Independent Samples: λ^2 test

λ^2 is a versatile non-parametric test. It can be used in a variety of situations and when measurements are nominal, ordinal or interval scale. The experimental hypothesis tested is whether two groups differ with respect to a certain characteristic. The distribution of the proportion of individuals in one group over various categories is compared with the proportion of cases in the other group. The null hypothesis to be tested is that these proportions do not differ. For an $r \times k$ contingency table, the df is $(r - 1)(k - 1)$, in which, **r** is the number of rows and k the number of columns.

For a 2×2 contingency table ($r = k = 2$), the following formula is more appropriate.

$\lambda^2 = \frac{n*(AD-BC-n/2)^2}{(A+B)+(C+D)+(A+C)+(B+D)}$, where A, B, C, D refer to cell frequencies.

Table 4.32 Table representing patients' response to different type of treatments

	Recovered	Sick	Total
Treatment 1	43	2	45
Treatment 2	17	7	24
Total	60	9	69

Two related samples

First check- is the data binary. **Use McNemar's test.**

Have more than two sets of data—**Use Cochran's Q test.**

For binary data, we use the **McNemar test**. This test is typically used in a repeated measures situation wherein each subject's response is elicited twice, once before and once after a specified event occurs. The McNemar test determines whether the initial response rate (before the event) equals the final response rate (after the event). This test is useful for detecting changes in responses due to experimental intervention in before–after designs. The same individual acts as his own control.

The measurement is either on a nominal or an ordinal scale. McNemar test is essentially a λ^2 test but uses only two cells, A and D, that represents a change in the response. The other two cells B and C do not come into the analysis at all. If $p(A)$ is the proportion of response under the A category and $p(D)$ the proportion of response under D category, then.

Null hypothesis: $H_0\ p_A = p_D = 1/2$.

Alternative hypothesis

H_1 $p_A \neq p_D$ for a two-tailed test and.

H_1 $p_A > p_D$ or $p_A < p_D$ for a one-tailed test. The McNemar test, a modified $\boldsymbol{\lambda}^2$ test, is given by

$$\lambda^2 = \sum_{AD} \frac{(f_0 - f_e)^2}{f_e} = \frac{\left(A - \frac{A+D}{2}\right)^2}{\frac{A+D}{2}} + \frac{\left(D - \frac{A+D}{2}\right)^2}{\frac{A+D}{2}}$$

When simplified, it becomes $\frac{(A-D)^2}{A+D}$, with 1 degree of freedom.

The approximation of a discrete distribution by a chi-square distribution is best achieved if a correction for continuity is applied. With correction for continuity, it will become $\frac{(|A-D|-1)^2}{A+D}$.

Case Study

A random sample of college students were asked if they supported the "Free Economy" move initiated by Dr. Manmohan Singh in 1991. The same group of students was re-interviewed in 2000 to see if the economic liberalization policy followed so far helped change their earlier stand on the issue. The data is cast in a 2 × 2 table. Have the changes in the intervening period helped change their earlier stand? (Table 4.33).

Suitable Tool: McNemar test.

Reason: The test is appropriate because the samples are related, and the data is on a nominal scale.

Solution

H_0: Proportion of students shifting from "Oppose" to "Support" p_A is the same as those shifting from "support" to "oppose" p_D.

That is $p_A = p_D$ that of "support".

H_1 $p_D > p_A$.

The critical point is that the "hard-core" people do not change their opinion easily. Those who are lesser extremists are susceptible to change their views considering the developments and their effects on the economy in the period of "take-off". So, the groups belonging to cells *A* and *D* of the above fourfold table are important for testing the significance of the change. The appropriate test is **McNemar's test of modified** λ^2 with correction for continuity. The test is appropriate because the samples are related and the data is on a nominal scale.

$$\lambda^2 = \frac{(|A - D| - 1)^2}{A + D} = \frac{(-21)^2}{60}$$

We select the 0.05 level for testing the significance of difference. From the table, $\lambda^2 \geq 6.64$ for degrees of freedom = 1 is significantly at 0.01 level for a one-tailed test. The probability of getting as large as 6.64 or more for a two-tailed test is $p <$ ½ (0.01). The observed λ^2 is in the critical region, and therefore, we reject H_0 and conclude that the students who opposed the issue at the beginning of its implementation changed their stand more than those who supported the policy at the start.

Cochran Q Test for *k*-Related Samples

Cochran *Q* test is an extension of the McNemar test for more than two samples. This procedure is used for simultaneous testing of differences among three or more (*k*) matched sets of sample frequencies or proportions. Matching may be achieved by selecting persons who are similar in respect to certain characteristics relevant to the study, as in randomised block design (RBD), or the same set of subjects used for different experimental conditions, as in Latin square design (LSD) or in time-series experiments with the same panel of subjects. A very simple and useful example where the Cochran *Q* test may appropriately be used is in item analysis for testing the number of items for *N* individuals is good for testing the differences in difficulty values. Data on success and failure on *k* number of items for *N* individuals are good for testing the differences in difficulty by this *Q* test. We can also test if responses of

Table 4.33 Responses towards 'free economy' in year 2000 and before 1991

		After 2000	
		Opposed (-)	Supported (+)
Before 1991	Supported (+)	20	70
	Opposed (-)	30	40

sum of squares to a particular item or issue differ under varying conditions. Voting behaviour is an excellent issue, which can be analysed by the Cochran Q test for changing the voting behaviour of a significant effect on the measured variable.

Interpretation of Cochran's Q test [12]

For applying the Q test, we need to arrange our data in a two-way table, the rows representing the Ss (N) and columns representing the treatments or items or time-anchor points (k). Consider an example of a group of students' successes and failures on several items of a test. 1 and 0 denote success and failure, respectively. Under H_0, "successes" and "failures" are randomly distributed both in rows and in columns of the table. Cochran has shown that if N (number of rows) is not too small.

$$Q = \frac{k(k-1)\sum_{i=1}^{N}\left(G_j - \overline{G}\right)^2}{k\sum_{i=1}^{N} L_i - \sum_{i=1}^{N} L_i^2}$$

Q is distributed approximately as chi-square with df $= k - 1$, where G_j is the total number of **"successes"** in the jth column. $\overline{G}$ is the mean of G_j, L_i is the total number of successes in the ith row. Since Q is approximated by the chi-square distribution, the probability associated under H_0 with as large a value as the observed Q can be ascertained from the table of chi-square with.

Degrees of freedom $= k - 1$.

If $Q \geq \lambda^2{}_{\alpha,k-1}$, we accept that the proportion or frequency of "successes" differs significantly among the various samples and therefore reject H_0 at that level. That is, the treatments have different impacts on measured variables.

Example 21 Compute Q from the following hypothetical data (Table 4.34).

By substituting the appropriate values in this equation, we have

$$Q = \frac{k(k-1)\sum_{i=1}^{N}\left(G_j - \overline{G}\right)^2}{k\sum_{i=1}^{N} L_i - \sum_{i=1}^{N} L_i^2} = \mathbf{0.998214}$$

For **3** df, the observed value of Q lies between the probability levels $0.80 < Q < 0.70$.

This probability is much smaller than the usual minimum acceptance level, $\alpha =$ 0.05. We, therefore, accept the null hypothesis H_0: the probability of "successes" is the same for all the four items against the alternative hypothesis H_1: the probabilities of "successes" in different items differ.

Conclusion: The difficulties of the items are almost the same.

Table 4.34 Cochran's Q test for different subjects

Subjects	Items				L_i	L_i^2
	I	II	III	IV		
A	1	0	1	1	3	9
B	1	0	1	1	3	9
C	0	1	1	0	2	4
D	0	1	0	1	2	4
E	1	1	0	0	2	4
F	0	0	0	1	1	1
G	1	0	1	1	3	9
H	1	0	1	1	3	9
I	0	1	0	0	1	1
J	0	0	0	0	0	0
K	0	1	0	1	2	4
L	1	1	1	0	3	9
M	1	0	1	0	2	4
N	0	0	1	1	2	4
O	1	1	1	1	4	16
P	1	0	0	1	2	4
Totals	9	7	9	10	35	91

Some Commonly Used Non-Parametric Tests for Ordinal Data

Case I

One sample—to test for randomness of the sample. Run test for both small and large samples.

Case II: Two Independent Samples

1. **Median test**—two independent samples are tested for their difference in central tendencies.
2. Mann–Whitney U test.

Case III: Two Related Samples

1. Sign test.
2. Wilcoxon matched pairs signed-rank test.

Case IV: k Independent Samples

1. Median test.

2. Kruskal–Wallis one-way ANOVA by ranks

Case V: k-Related Samples

1. Friedman two-way ANOVA by ranks.

Note: Chi-square tests for independence, Fisher's exact test and randomization test of independence

All three tests in this section deal with nominal variables and are used to test if proportions are the same in different groups. When the study consists of more than two nominal variables, we use the chi-square test of independence. When the study consists of only nominal variables, we use Fisher's exact test or randomization test of independence. The table below summarizes the three tests (Table 4.35).

IMPORTANT: If the samples are not independent, but instead are before-and-after observations on the same individuals, the researcher should use McNemar's test.

One Sample: Wald–Wolfowitz's runs test or the test of Randomness

Wald–Wolfowitz is an American Jewish Statistician, who has made immense contributions to the field of statistical decision theory, information theory and non-parametric statistics.

The run test is a very useful device for testing randomness. We study a set of observations or items of a sample in the order in which they occur and see if there exists any definite pattern in their occurrence or if the scores or items appear unsystematically.

One run is defined as a succession of identical symbols followed or preceded by different symbols

A simple example is the tossing of a coin. We write 1 for the appearance of "Head" and 0 for "Tail". That is, 1 and 0 or + and – or any other symbols that represent a series of successive tosses. A succession of 20 tosses is represented by a set of numbers or symbols as follows:

1 0 0 1 1 1 0 1 0 0 0 1 1 0 1 1 1 1 0 1.

For a continuous run of zero's and one's in this series, the runs are calculated as follows.

Table 4.35 Different tests to be used at different situations

Situation	Test
Large sample sizes (greater than 1000)	Chi-square test of independence
Small sample sizes (less than 1000)	Fisher's exact test
Small sample sizes (less than 1000) and large numbers of categories	Randomization test of independence

$$\underline{1}\ \underline{0\,0}\ \underline{1\,1\,1}\ \underline{0}\ \underline{1}\ \underline{0\,0\,0}\ \underline{1\,1}\ \underline{0}\ \underline{1\,1\,1\,1}\ \underline{0}\ \underline{1}$$

1 2 3 4 5 6 7 8 9 10 11

or

$$\underline{+}\ \underline{--}\ \underline{+++}\ \underline{-}\ \underline{+}\ \underline{---}\ \underline{++}\ \underline{-}\ \underline{++++}\ \underline{-}\ \underline{+}$$

1 2 3 4 5 6 7 8 9 10 11

This sample of observations contains a random set of 1 s and 0 s. It begins with a run of 1 "Positive" or "Plus" followed by a run of 2 **negatives or minus**. Another run of 3 positives appears next. We underline each run separately and number them. In the hypothetical toss, we have as many as 11 runs ($r = 11$). It is this number of runs for a sample that indicates the degree of randomness. With a sample of 20 observations, we can have 20 runs at the most, each run consisting of only one symbol preceded and succeeded by different symbols. Example: $+ - + - \ldots$ or $1\ 0\ 1\ 0 \ldots$, etc., 20 runs for $n = 20$ is too many indicating a regular pattern or indicating lack of independence in either of the observations. As the number of tosses increases, the proportion of **"Heads"** and **"Tails"** will tend towards a value in the neighbourhood of 0.5. A X^2 test or binomial test will yield results from which we shall have no reasons to suspect the fairness of the coins. But a toss of a "fair coin" would not only produce "Heads" and "Tails" in any systematic manner, their appearance should also display randomness. This is what is tested by the run test.

Run test for small sample

Let n_1 be the number of elements of one kind and n_2 be the number of elements of the other kind so that the sample of size $n_1 + n_2 = n = 20$ will be considered a small sample. For $n_1 = 10$ and $n_2 = 10$, the number of runs (r) less than or equal to 6 or greater than or equal to 16 is significant at 0.05 level of significance leading us to the question of the randomness of the sample, thereby leading to the rejection of the null hypothesis H_0. The critical values of the number of runs (r) are obtained from tables. The null hypothesis tested is that the two types of events occur in random order. The alternative hypothesis is that the events occur in a predetermined order.

The split between the two events need not be equal, it is not necessary that be n_1 be equal to n_2 Thus, for an 18-item sample of 12–6 split, the two limits of r (the number of runs) are 4 and 13. That is, $r \leq 4$ or $r \geq 13$ indicate non-randomness. In the 20-item example, if $n_1 = n_2 = 10$. The two tabulated values are 6 and 16, which means that at 0.05 level of confidence, a random sample shall contain more than 6 and less than 16 runs. ($6 < r < 16$) Values of $r = 6$ and $r = 16$ are in the critical region and the runs so obtained cannot be explained by the chance factor. Our observed values of runs (above), $r = 11$ lies between the two tabulated and therefore we accept the null hypothesis and conclude that the order in which "Head" and "Tail" appears is generally random.

Run test for large sample

For large samples, when we have n_1 objects of one kind (say Head) and n_2 objects of another kind (say Tail) arranged along a line, the number of possible arrangements is $C(n_1 + n_2, n_1)$ where $n_1 \leq n_2$. The number of arrangements with exactly r runs is given by

$f_r = 2C(n_1 - 1, k - 1)C(n_2 - 1, k - 1)$ where $2k = r$ (r is even) and when r is odd ($r = 2k - 1$). Here, k can take all integral values between 1 and $n_1 - 1$. The probability that the number of runs $r < r'$ which is an estimated value in a random arrangement is given by

$$P(r < r') = \frac{\sum_{r=2}^{r'} f_r}{c(n_1 + n_2, n_1)}$$

Although the probability of obtaining a particular number of runs for higher values of n_1 and n_2 can be found out by using the above formula, the calculations become tedious especially if the number of runs is not too small. When either n_1 or n_2 is larger than 20, the distribution of r is approximately normal with mean, $\mu_r = \frac{2n_1n_2}{n_1+n_2} + 1$ and variance, $\sigma_r^2 = \frac{2n_1n_2(2n_1n_2-n_1-n_2)}{(n_1+n_2)^2(n_1+n_2-1)}$.

For a large one-sample set of observations, H_0 may be tested by the z-test, where

$$Z = \frac{r - \mu_r}{\sigma_r} \sim N(0, 1)$$

Example 22 In an industrial production line, items are inspected periodically for defects. The following is a sequence of defective items denoted as **D** and non-defective items denoted as **N**, produced by the production line:

D D N N N D N N D D N N N N N D D D N N D N N N N D N D.

Use the large sample theory for the runs test, with a significance level of 0.05, to determine whether the defective entities are occurring at random.

Suitable Tool: One sample runs test for large sample.

Reason: The null hypothesis refers to the randomness of a single group of events.

Solution

Null Hypothesis H$_0$: The defectives are occurring at random.

Alternative Hypothesis H$_1$: The defectives are not occurring at random.

Test Statistic: $Z = \frac{r-\mu_r}{\sigma_r} \sim N(0,1)$.

Level of significance: $\alpha = 0.05$.

Calculations

$$\underline{DD}\ \underline{NNN}\ \underline{D}\ \underline{NN}\ \underline{DD}\ \underline{NNNN}\ \underline{DDD}\ \underline{NN}\ \underline{D}\ \underline{NNNN}\ \underline{D}\ \underline{N}\ \underline{D}.$$
$$1\quad 2\quad 3\ 4\quad 5\quad 6\quad 7\quad 8\quad 9\quad 10\quad 11\ 12\ 13$$

Runs, $r = 13, n_1 = 11, n_2 = 17.$

Mean, $\mu_r = \frac{2n_1n_2}{n_1+n_2} + 1 = \frac{2*11*17}{11+17} = \mathbf{14.36}.$

Variance, $\sigma_r^2 = \frac{2n_1n_2(2n_1n_2-n_1-n_2)}{(n_1+n_2)^2(n_1+n_2-1)} = \mathbf{6.113}.$

Test Statistic: $|Z| = \left|\frac{r-\mu_r}{\sigma_r}\right| = \left|\frac{13-14.36}{\sqrt{6.113}}\right| = |-0.55| = \mathbf{0.55}$

$$Z_{\alpha/2} = Z_{0.025} = 1.96$$

Critical region: $|Z| \geq Z_{\alpha/2}$.

Conclusion: Since the obtained value of z (0.55) lies outside the critical region, we accept H_0 and conclude that the defects are occurring at random.

Mood's Median Test

Answers a very simple question, "Are the Medians of the populations from which two or more samples are drawn identical?" This actually is an extension of **Pearson's chi-squared test**. The median score for the combined k-samples is first determined. The score of each sample is then given one of the two signs, **"+"** if the score exceeds the median of the pooled distribution or **"−"** if the score is smaller than the median score. That is, we prepare a $2 \times k$ table based on a median split. Each column, representing samples, has frequencies for both **"+"** and **"−" signs**. The null hypothesis tested is that all the k samples come from the same population and therefore the split of each between **"+"** and **"−"** signs is the same. The test used for testing the significance of difference in splits is chi-square with **df = k − 1.** This extended median test is less powerful than the Kruskal–Wallis test, to be discussed shortly. Much of the significance of a score is lost when we simply dichotomize it on the basis of a common median. When we convert a set of observations into ranks, as we do in the Kruskal–Wallis test, much information of the original observation is retained in ranks.

Median Test—Two Independent Samples Median test can be regarded as a special case of both Fisher's exact probability and chi-square test for a 2×2 contingency table. Here, two independent samples are tested for their differences in their central tendency, the median. A more appropriate statement would be if the two samples have been drawn from the same population. The procedure is simple. We calculate the median of the composite group. Divide the two samples separately into those whose scores lie above the pooled median and those whose scores lie below the median. It takes the following form (Table 4.36).

When both the samples are drawn from the same population, the median that is computed separately for the two samples shall be the same as the median for the combined group except for marginal error due to sampling fluctuations. This would mean that 50% of individuals of either group would be located in the **"above-median category"**. That is, the proportion of individuals in cells A and B or cells C and D shall be equal.

If n_1 and n_2 are small and if some of the observed cell frequencies are less than 5, Fisher's exact probability test may be applied.

- If n_1 and n_2 are sufficiently large and no cell frequency is less than 5, we may use X^2 test with degrees of freedom 1. Generally, when $\mathbf{n_1 + n_2 > 40}$, we use X^2 test with correction for continuity.
- If $n_1 + n_2$ is between 20 and 40 and no cell frequency is less than 5, X^2 with continuity may also be applied.
- When $\mathbf{20 < n_1 + n_2 < 40, n_1 + n_2 < 20}$ and some cells have less than 5 observations, Fisher's test is to be used.

Example 23 Below are the age distributions of two samples drawn from the same population. Test if they really are random samples drawn from the same population (Table 4.37).

Suitable Tool: X^2 test with correction for continuity.

Table 4.36 Scores of individuals categorised as above and below median

No. of individuals whose score is	Sample 1	Sample 2	Total
Above the median	A	B	$A + B$
Below median	C	D	$C + D$
Total	$\mathbf{A + C = n_1}$	$\mathbf{B + D = n_2}$	$\mathbf{n_1 + n_2 = n}$

Table 4.37 Age distributions from two samples from same population

Age	Sample 1	Sample 2	Combined	Cumulative frequencies
15–19	5	10	15	15
20–24	10	13	23	38
25–29	20	25	45	83
30–34	13	10	23	106
35–39	2	2	4	110
Total	$\mathbf{N_1 = 50}$	$\mathbf{N_2 = 60}$	$\mathbf{N_1 + N_2 = 11}$	

Reason: None of the cell frequencies is less than 5 and the total sample size is appreciably large.

Solution
$M_t = \left(l + \frac{\frac{n}{2} - F}{f_m}\right) * c = 24.5 + \frac{55-38}{45} * 0.5 = \mathbf{26.39}.$

In sample l, the number of observations that are ≤ 26.39 is $15 + \frac{26.39-24.5}{5} * 0.20 = 22.55 \leq 23$.

The same in sample 2 is $23 + \frac{1.89}{5} * 0.25 = 32.45 \geq 32$ (Table 4.38).

With this information, we construct a 2×2 contingency table with a median split.

$\mathbf{H_0}$ The proportion of individuals in cells A and B or C and D is the same.
$\mathbf{H_1}$ The proportion of individuals in cells A and B or C and D are different.

Since none of the cell frequencies is less than 5 and the total sample size is appreciably large, the appropriate test for testing the significance of difference in cell frequencies is chi-square with correction for continuity. It will be a two-tailed test and the level of significance selected is 0.05.

Computations

$$X^2 = \frac{n\left(|AD - BC| - \frac{n}{2}\right)^2}{(A+B)+(B+D)+(A+C)+(C+D)}$$
$$= \frac{110(|264 - 644| - 55)^2}{55 * 50 * 55 * 60} = \mathbf{0.33}$$

Tabulated value: $X^2_{0.05,1} = \mathbf{3.84}.$
Critical region: $X^2 < X^2_{0.05,1}$.
Since obtained value X^2 is far less than the tabulated value 3.84, we fail to reject H_0 at alpha 0.05 and conclude that there is no sufficient evidence that two samples are drawn not from the same population.

Mann–Whitney U test

Mann–Whitney U test is applicable to two-sample independently drawn in which at least ordinal level measurement has been achieved. This is a very powerful non-parametric test, often regarded as the best alternative to a **t-test.** At times when data does not satisfy the assumptions under **t-test** or if the measurement is not strictly in an interval scale, then Mann–Whitney U test is best suited.

This method is commonly used during opinion surveys, judgements, merit ratings etc. The null hypothesis tested in the **U-test** is that populations A and B have the same

Table 4.38 2×2 contingency table with a median split for sample 1 and 2

	Sample 1	Sample 2	Total
Above median	27 (A)	28 (B)	55 ($A + B$)
Below median	23 (C)	32 (D)	55 ($C + D$)
Total	**50 (A + C)**	**60 (B + D)**	**110**

distribution. The alternative hypothesis is that population *A*/*B* is stochastically larger than *B*/*A*.

The Mann–Whitney U test is used to compare whether there is a difference in the dependent variable for two independent groups. It compares whether the distribution of the dependent variable is the same for the two groups and therefore from the same population.

Mann–Whitney U test doesn't compare the medians—This is a popular saying you will usually come across, which means that this test does not compare medians of 2 groups of data, whereas it compares the mean ranks of two groups of data. Figure 4.22 explains better through infographics. Mann–Whitney tests are not related to the medians nor the distribution of the data as they consider only the ranks of data (Fig. 4.23).

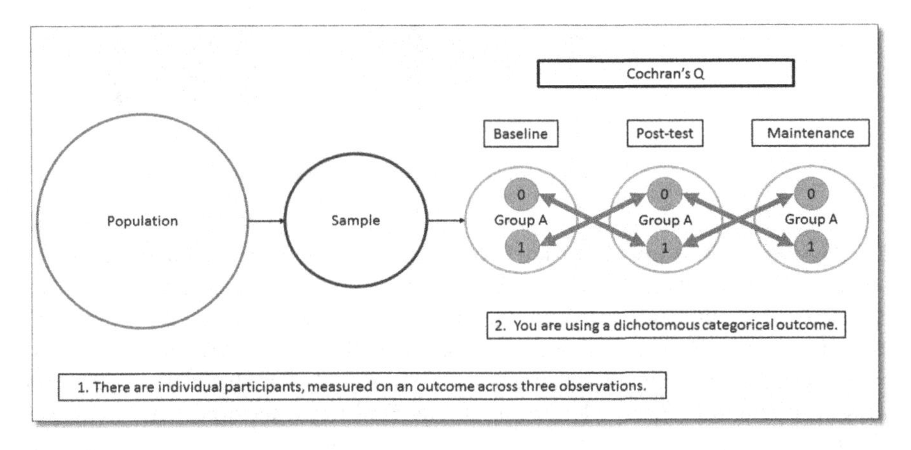

Fig. 4.22 Diagrammatic representation of Cochran's Q test

Fig. 4.23 Wald-Wolfowitz

Fig. 4.24 Mann Whitney U test

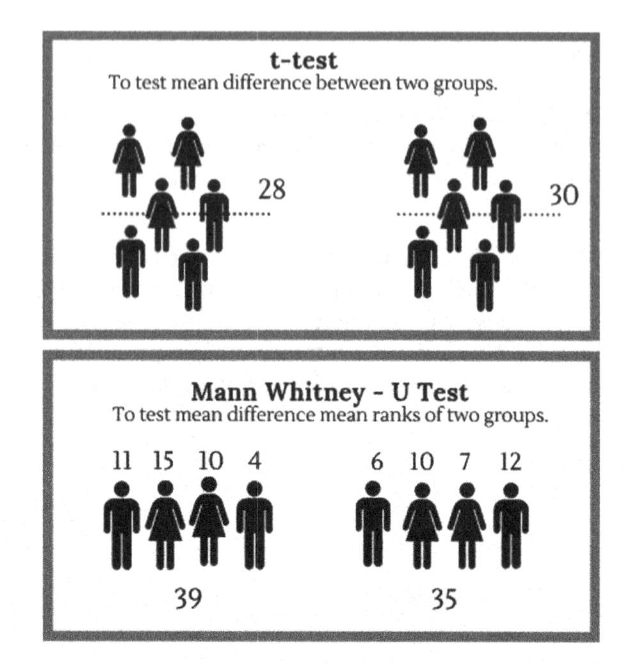

The logic behind Mann–Whitney U test: This test ranks the data under different criteria and then tests to see how different they are, by using its average. If the two data sets are significantly different, then most of the higher ranks are concentrated towards one criterion and the lower ranks belong to the other criteria. In case the two criteria are similar to each other, the ranks are uniformly distributed across the data with negligible differences. Therefore, Mann–**Whitney Utest statistic** reveals the difference in ranks. The lesser the value of U, the lesser the possibility of occurrence by chance. The hypothesis, thus formulated will be (Fig. 4.24)

Null Hypothesis H_0: There is no difference in the sum of the rankings in the two groups in the population.

Alternate Hypothesis H_1: There is a difference in the sum of the rankings in the two groups in the population.

Fors the Mann–Whitney U test, the null hypothesis is that *t* The alternative hypothesis is that there is a difference in the sum of the ranks.

H_1 will be accepted if $P\ (a > b) < 1/2$. A two-tailed test could also be applied if the prediction of differences does not state the direction so that H_1: $(a > b) \neq 1/2$.

The procedure for the U test is stated under.

1. Combine the two samples into one and arrange the scores in order of magnitude, lowest to highest.

An example

See Table 4.39.

Table 4.39 Mann Whitney U test for Group A and group B

Group A	10	13	17	19		25	$n = 5$					
Group B	7	9	18	20		22	24	$n = 6$				
Combined	7	9	10	13		17	18	19	20	22	24	25
	B	*B*	*A*	*A*		*A*	*B*	*A*	*B*	*B*	*B*	*A*

2. Mark the observations on the ordered scale by writing *A* or *B* depending on if they come from sample *A* or sample *B*. So, the identity of the observations is preserved.
3. Usually, the *B* group is our control group. Count the number of A that immediately precedes each of *B*. Thus, for *B*-7 and also for *B*-9 there are no *A* scores preceding them. The next *B* score *B*-18 has 3 *A* observations preceding it. So, we write 0 for *B*-7, 0 for *B*-9 and 3 for *B*-18. *B*-20 is preceded by 4 *A* observations *A*-19, *A*-17, *A*-13, *A*-10 so also for *B*-22 and *B*-24. The sum of these scores is the required **U value**. In this case, as explained

$$U = 0 + 0 + 3 + 4 + 4 + 4 = \mathbf{15}$$

The process

- This gives us the number of occasions when sample *A* observation precedes sample *B* observations.
- Tabulated values for the sampling distribution of U for different values of n_1 and n_2 are available. From this, we can determine the probability of obtaining any U value as extreme as the observed one.
- From the U test statistical table, we get, $n_1 = 5$ and $n_2 = 6$, the probability of obtaining U under H_0 equal to or less than 15 is **0.535** which is much more than the probability level 0.05 we generally require for such data.
- U values $= 5$ are required for U to become significant at 0.05 level. We, therefore, reject H_1.

Computation of U when n_2 is large, between 9 and 20

Here we use a separate U table for values of n_2 between 9 and 20. This table gives us critical values of U for selected levels of significance $\alpha = 1$, 10 and 5% for a one-tailed test. For a two-tailed test, the corresponding critical values are $\alpha = 0.5\%$, 5% and 2.5%, respectively. If the observed value of U is less than or equal to the critical value (tabulated value), H_0 may be rejected at the stated level of significance indicated on the U table.

Follow the same procedure for computing *u*-test statistic, which is described earlier. Suppose $n_1 = 7$ and $n_2 = 15$ and $U = 21$. For a one-tailed test, the critical value is 10 at 0.001 level, 19 at 0.01 level, 24 at 0.025 level and 28 at 0.05 level of significance.

The obtained U = 21 enables us to reject H_0 both at 0.025 and at 0.05 levels for a one-tailed test.

Alternative computations for the value of U for samples between 9 and 20.

The above procedure for counting U will be tedious if the two samples are fairly large. The following fairly simple and straightforward method can be used to obtain an identical result. Let the sample sizes of Groups C and D be 15 and 12, respectively. Combine the two groups (keeping their identity intact) and assign rank 1 to the smallest score of the combined series, rank 2 to the next lowest score and so on. U is given by

$$U = n_1 n_2 + \frac{n_1(n_1 + 1)}{2} - R_1$$

Or

$$U = n_1 n_2 + \frac{n_2(n_2 + 1)}{2} - R_2$$

where R_1 is the sum of ranks assigned to the group having a sample size of n_1 and R_2 is the sum of ranks assigned to the group having a sample size of n_2 (Table 4.40).

$$U = n_1 n_2 + \frac{n_1(n_1 + 1)}{2} - R_1 = 15 * 12 + \frac{15 * (15 + 1)}{2} - 255 = 45$$

$$\mathrm{U} = n_1 n_2 + \frac{n_2(n_2 + 1)}{2} - R_2 = 15(12) + \frac{12(12 + 1)}{2} - 123 = 135$$

The two values of U are different. Here we consider the smallest value of U which is 45 in this example. There are two checks in the computation. The first is the sum of ranks. The sum of the first ***n*** natural numbers is $\frac{n(n+1)}{2}$, and, hence the sum of the ranks of our example is $\frac{27(27+1)}{2} = 378$.

Since the ranks are natural numbers from 1 to 27, which checks the sum of the ranks $R_1 + R_2 = 255 + 123 = \mathbf{378}$. The second check is for U. The researcher must check if he/she has found U' (larger value) by the following transformation.

$$U = n_1 n_2 - U' = (15 * 12) - 135 = 45$$

The smaller of the two values of U is the one whose sampling distribution is the basis of the tabulated values. **Both counting method and rank method would give the same result.**

We shall work it out with the data for which U was determined by **the counting method** (Table 4.41).

$$U = 5 \times 6 + (5 \times 6)/2 - 30 = 15$$

Table 4.40 Scores and ranks of group C and D

C group		D group	
Scores	Ranks	Scores	Ranks
95	27	60	15
55	11	42	4
65	18	46	5
75	23	53	9
80	25	39	2
50	7	64	17
48	6	51	7
59	14	58	13
63	16	37	1
82	26	70	20
56	12	54	10
74	22	69	19
73	21		
41	3		
79	24		
	$R_1 = 255$		$R_2 = 123$

$$U = 5 \times 6 + (6 \times 7)/2 - 36 = 15$$

Both U values agree with the U obtained by counting method. If the observed value of U is less than or equal to U_α, then H_0 is rejected. The critical values U_α for $n_1 = 12$ and $n_2 = 15$ at 0.001, 0.01 and 0.05 levels are 28, 42 and 49, respectively. $U = 45$ is less than 49, and therefore, we reject H_0 at 0.05 level but not at 0.01 nor at 0.001 level .

Table 4.41 Mann Whitney U test for scores of group C and group D

A scores	Rank	*B* scores	Rank
10	3	7	1
13	4	9	2
17	5	18	6
19	7	20	8
25	11	22	9
		24	10
	RA = 30		RB = 36

Computation of U value when n_2 is larger than 20

Mann–Whitney have shown that as the sample size increases, the sampling distribution of U fast approaches the normal distribution such that $U_\mu = \frac{n_1 n_2}{2}$ and $\sigma_\mu = \sqrt{\frac{n_1 n_2 (n_1 + n_2 + 1)}{12}}$

When $n_2 > 20$, the significance level of the observed value of U can be obtained by the

Test statistic: $Z = \frac{U - U_\mu}{\sigma_\mu} \sim N(0,1)$.

Decision Criteria: H_0 is rejected if $Z \geq Z_\alpha$

Example 24 Given below are the mileages (in thousands of miles) of two samples of automobile tyres of different brands, say I and II before they wear out.

Tire I : 34 32 37 35 42 43 47 58 59 62 69 71 78 84
Tire II : 39 48 54 65 70 76 87 90 111 118 126 127

Use Mann–Whitney test to check if tyre II gives more median mileage than tyre I.

Use $\alpha = 0.05$.

Suitable Tool: Mann–Whitney test.

Reason: We are comparing two samples that are independently drawn.

Solution
Null hypothesis H_0: The median mileages of the two brands of tyres are the same.

Alternative hypothesis H_1: The median mileage of tyre I is less than that of tyre II.

Test statistic: $Z = \frac{U - U_\mu}{\sigma_\mu} \sim N(0,1)$.

Level of significance: $\alpha = 0.05$.

Calculations
See Table 4.42.

$$R_1 = 138, n_1 = 14, n_2 = 12$$

$$U = n_1 n_2 + \frac{n_1(n_1 + 1)}{2} - R_1$$

$$14(12) + \frac{14(14 + 1)}{2} - 138 = 135$$

$$U_\mu = \frac{n_1 n_2}{2} = 84 \text{ and } \sigma_\mu = \sqrt{\frac{n_1 n_2 (n_1 + n_2 + 1)}{12}} = 19.44$$

Table 4.42 Mileages and ranks of Tire I and Tire II

Tyre I mileages	Ranks	Tyre II mileages	Ranks
34	2	39	5
32	1	48	9
37	4	54	10
35	3	65	14
42	6	70	16
43	7	76	18
47	8	87	21
58	11	90	22
59	12	111	23
62	13	118	24
69	15	126	25
71	17	127	26
78	19		
84	20		
	$R_1 = 138$		

$$Z = \frac{U - U_\mu}{\sigma_\mu} \sim N(0, 1) = \frac{135 - 84}{19.44} = 2.623$$

$$Z_\alpha = 2.58$$

Critical region: $Z \geq Z_\alpha$.
Since the obtained value of z 2.623 lies in the critical region, we reject H_0.

Conclusion: The median mileage of tyre I is less than that of tyre II.
Two related samples: Sign Test

The sign test gets its name from the fact that it uses **"plus"** and **"minus"** signs as its data and not the absolute values of measurement. Only the signs of the difference between two paired observations are used. Since the sign test uses only a part of the information provided by the basic data, it cannot be considered as powerful a test like the t-test or u-test. However, the sign test has some advantages over parametric statistics. That is, it neither makes any assumptions relating to the form and shape of the distribution nor does it assume that all the responses are drawn from the same population. The only restriction is that the measurements must at least be on an ordinal scale.

In the selection of pairs, however, one has to see that they are matched at least in respect to those extraneous conditions, which are relevant to the study objectives. This is best achieved if each subject acts as its own counterpart in the matched pair.

Let X_b be the score of subject B and X_a the score of subject A in the pair AB in the treatment condition in the "before–after" design. Then X_a and X_b are the scores of the matched pairs. The null hypothesis suggested in the sign test is,

$$P(\chi_a > \chi_b) = P(\chi_a < \chi_b) = 1/2$$

That is, if there were n pairs of observations, under H_0 there would be as many numbers of $X_a > X_b$ as there are numbers $X_a < X_{b.}$ This is the same as saying that if H_0 were true, then half of the difference between the pairs would be positive and the other half negative. H_0 is rejected; only a few of a particular sign (+ or −) are observed in the set of differences.

We make use of the binomial expansion to find out the probability associated with the occurrence of a particular number of positives and negatives, putting $p = q = 0.5$, n being the number of pairs. It often happens that some pairs show neither positive nor negative sign as difference, $X_{\text{at}} = X_{\text{bt}}$. In such situations, these pairs are to be dropped from the list and n adjusted accordingly. If there are r number of pairs showing no difference, our final analysis will be carried out based on $(n - r)$ pairs of difference.

Case Study

Opinions of 20 parents were sought on the severity of punishment to be given to their child if he/she happened to tamper with the school progress report. The ratings on the severity of punishment were on a 5-point rating scale (4 for highest level of punishment, 0 for least). Do father's (Male) and mother's (Female) opinions differ?

Suitable Tool: Sign Test.

Reason: The measurements are on a partially ordered scale (a few zero differences). Each couple may be regarded as a matched pair since the punishment was to be given to their children. Two samples are therefore related. The data being in partially ordered scale, sign test seems appropriate for testing of difference (Table 4.43).

Solution

H_0 The number of positive signs = Number of negative signs.
H_1 The number of positive signs is larger than the number of negative signs. That is, fathers are more strict than mothers.

The sampling distribution of M < F under H_0 by binomial distribution is $p = q = 0.5$.

The data yield indication as to the direction of the difference. Therefore, the rejection region is one-tailed. We shall accept H_0 if probability under H_0 of M < F as large as 6 is equal to or less than $\alpha = 0.05$.

There are two cases of **"tie"**. We drop these pairs and adjust our n from 20 to 18.

We can calculate the probability of obtaining M < F = 6 directly from binomial expansion or from the table, which gives probabilities associated with values as small as the observed values of x in the binomial test. From the table, we find that for $x = 6$ and $n = 18$, the probability of obtaining $x = 6$ is 0.119. This value is much

Table 4.43 Opinions of parents on severity of punishments

Couple	Rating by M	Riling by F	Direction of difference	Sign	Couple	Rating by M	Riling by F	Direction of difference	Sign
1	4	2	M > F	+	11	1	2	M < F	–
2	3	3	M = F	0	12	3	1	M > F	+
3	2	1	M > F	+	13	4	3	M > F	+
4	3	1	M > F	+	14	3	2	M > F	+
5	2	3	M < F	–	15	2	1	M > F	+
6	1	2	M < F	–	16	3	1	M > F	+
7	4	2	M > F	+	17	4	1	M > F	+
8	2	4	M < F	–	18	3	2	M > F	+
9	1	1	M = F	0	19	1	3	M < F	–
10	3	2	M > F	+	20	2	3	M < F	–

larger than $\alpha = 0.05$, and therefore, we accept H_0. That is, the opinion of fathers and mothers do not differ significantly in matters of disciplinary measures against their children.

Case Study

A survey was conducted on students of a university department on the time they spent on watching the Singer Cup competition among four nations. Their classes stratified the sample. Would you say that the stratified samples belong to the same population based on the data furnished below? Time spent (hours) watching the telecast (Table 4.44).

Suitable Tool: Non-parametric extended median test for testing the hypothesis.

Reason: There are $k = 5$ samples drawn from different classes. The report on time spent is only an approximate value. This is a case of one-way analysis of variance for simultaneous testing of differences between samples. Considering the limitations in the measurement and to avoid the assumptions related to the F-test, we choose a non-parametric extended median test for testing the hypothesis.

Solution

1. **Null Hypothesis, H_0**: there is no difference between the samples in the average hours spent on attending the sports telecast.
   Alternative Hypothesis, H_1: There is a difference between the samples in the hours spent on the telecast.
2. **The rejection region**: consists of all values of chi-square with df = 45 greater than or equal to the tabulated value $X^2_{0.05,4}$
3. **Calculations** (Table 4.45).

i. The median hour of the combined distribution is found to be 6.69. Each sample is divided into two categories—below and above median based on this value. A table composed of 5 × 2 cells is prepared. The expected frequency for each cell

Table 4.44 Time spent on watching Singer Cup competition among four nations

Undergraduate			Postgraduate		Combined	
1st	2nd	3rd	1st	2nd	Hours	Frequency
30	10	5	38	8	0–4	14
25	35	2	15	6	5–9	8
4	20	0	3	12	10–14	3
2	0	0	1	5	15–19	4
0	10	15	4	0	20–24	2
16	5	20	7	0	25–29	1
6	7		0	3	30–34	1
18					35–39	2

Table 4.45 Mann Whitney U test for hours spent on the telecast

	Undergraduate			Postgraduate		Total
	First	Second	Third	First	Second	
Below median	4 (4.34)	2 (3.8)	4 (3.26)	4(4.38)	5(3.8)	19
Above median	4 (3.66)	5 (3.2)	2 (2.74)	3(3.2)	2 (3.2)	16
Total	8	7	6	7	7	35

is calculated by the formula $f_{eij} = \frac{r_i k_j}{n}$ where $n = 35$ is the sum of all samples. The expected frequencies are shown in parentheses against f_0 in the same cell.

ii. X^2 is calculated by the formula $\sum \frac{(f_o - f_e)^2}{f_e}$. The calculated value is X^2 =3.146.

4. **Decision:** The probability of occurrence under H_0 of a value as large as $\chi^2 =$ 3.146, df $= 4$ is greater than 0.50. Therefore, we accept H_0 and conclude that the samples do not differ significantly on the average time spent on attending a special sports programme.

Some Non-parametric test for relationship between variables

Case I

When measurements are in nominal scale.

1. Wilk's λ.
2. Coefficient of contingency.
3. Cramer's V.
4. Phi coefficient.

Case II: Measurement is an Ordinal Scale

1. Spearman's rank correlation.
2. Kendall's rank correlation.

Case III

k sets of ranks—Kendall's coefficient of concordance.

Case Study

25 students were interviewed to find out if the new rules of the library were fair enough. 20 were in favour. Does the data suggests that the new rules are acceptable to students?

Suitable tool: Normal expansion to binomial.

Reason: Data is in discrete categories. A single sample has been chosen. $p = 20/25 = 0.80$, this is not very small.

Scenario: A public opinion poll in a town was interested in the type of content that adults in the age bracket of 20–50 years prefer to watch on TV. A random sample of 120 was selected and asked, "Given your preference, would you prefer to watch serials (S), or sports (R) or news (N) on TV?" Of the respondents, 40 indicated a preference for serials, 50 selected sports, and 30 selected news. Can the researcher conclude that the proportion of people in the population preferring to watch the three contents is almost the same?

Suitable tool: Chi-square test.

Reason: Data is in multiple discrete categories and is drawn from a single sample. The research question needs a comparison between observed and expected frequencies.

Scenario: A random sample of PU college students was asked if they supported "internal assessment" in 2005. The same sets of students, now in degree, were asked the same question if this concept helped them to get more marks, thus changing their views in 2008. Have the changes in the intervening years changed their view?

Suitable tool: McNemar's test of modified chi-square.

Reason: The people who were opposed and later changed their minds as well as those who were in favour of and then changed their opinions are important here. They help in determining the significance of change. The samples are related and the data is on a nominal scale. Thus, McNemar's test is appropriate here.

McNemar's test of modified chi-square

McNemar's test was created by **Quinn McNemar**, a professor in the psychology and statistics department at Stanford University and was first published in a **Psychometrika article in 1947**. This test is also called, **"Within-subjects chi-squared test"** It is used to compare two population proportions that are related or correlated to each other. This test is also used when we analyse a study in which two subjects are tested before-and-after time periods. It is applied by a 2 × 2 contingency table with the dichotomous variable. It is also known as the **"Test for marginal homogeneity for K x K table"** (Fig. 4.25).

Case Study

A researcher randomly selects 80 parents to get their opinion on the introduction of "sex education" classes in high school. After taking their opinion, they were given material highlighting the importance of such classes to children. A survey was conducted again, and the results were tabulated in the form of the table given below (Table 4.46).

Can the researcher conclude whether there was a significant change in the attitude of parents after reading the given material?

Suitable tool: This requires McNemar's test as.

(i) The subjects are randomly drawn.
(ii) The collected data can be grouped into "before–after" categories.

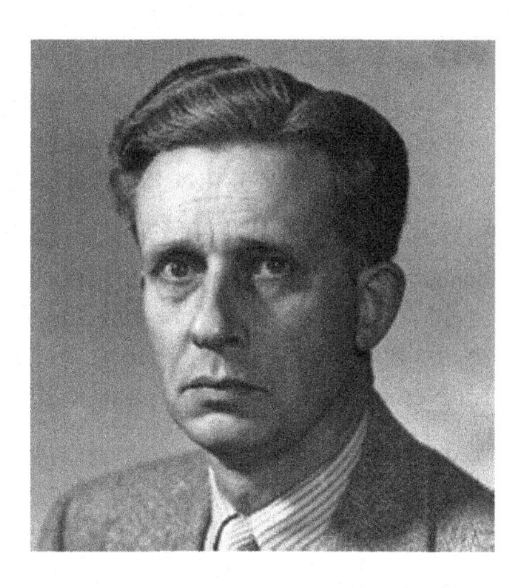

Fig. 4.25 Quinn McNemar

Table 4.46 Opinions of parents before and after reading the material Quinn McNemar

Opinion before reading material		Opinion after reading material	
		Favour of classes	Not in favour of classes
	Favour of classes	30	14
	Not in favour of classes	34	02

(iii) Data is nominal.
(iv) No assumptions are made about normality, continuity or equality/ in-equality of variances.

Solution
Step 1: Setting up the null and alternative hypothesis.

$\mathbf{H_0}$ There is no significant change in the attitude of parents after reading the given material.
$\mathbf{H_1}$ There is a significant change in the attitude of parents after reading the given material.

Caution: In such tests, we need to find expected frequencies first. If they turn out to be less than 5, the binomial test has to be used rather than McNemar's test. If they are equal to or more than 5, we can safely use McNemar's test.

What happens when the wrong tool is used, based on the above example?

Wrong tool 1: In the above example, some researchers may take $\widehat{p_1} = 30/80$ and $\widehat{p_2} = 14/80$ and use the z-test for proportions. This is incorrect as a valid p-value is not obtained. The same sample of 80 people is used in each proportion and it does not compare independent samples.

Wrong tool 2: Some researchers may perform a chi-square or Fisher's exact test on the above 2 × 2 table: These tests would test whether the proportion of those in favours among parents has a relation to the proportion of children who are in favour and will not test whether the proportions of opinions are different or are the same.

Scenario: A medical researcher wants to determine whether or not a particular drug has an effect on a disease. He records a count of the individuals as + and – sign (or 0 and 1) for denoting before and after being given the drug. McNemar's test is applied to test if the drug has an effect on the disease. This is because the data is paired and McNemar's test is, in fact, a paired version of the chi-square test.

Note 1: If an analyst is comparing proportions between the two groups or testing for independence between two categorical variables then the appropriate tool is the chi-square test.

Note 2: If more than 20% of the expected cell frequencies are less than 5, then the researcher should use Fisher's exact test.

Case Study

McDonald and Kreitman (1991) sequenced the alcohol dehydrogenase gene in several individuals of three species of Drosophila. Varying sites were classified as synonymous (the nucleotide variation does not change an amino acid) or amino acid replacements and they were also classified as polymorphic (varying within a species) or fixed differences between species. The two nominal variables are thus synonymicity ("synonymous" or "replacement") and fixity ("polymorphic" or "fixed"). In the absence of natural selection, the ratio of synonymous to replacement sites should be the same for polymorphisms and fixed differences. There were 43 synonymous polymorphisms, 2 replacement polymorphisms, 17 synonymous fixed differences, and 7 replacement fixed differences (Table 4.47).

Suitable tool: Fisher's exact test.

Wrong tool: Chi-square test.

Reason: Since the expected numbers in some classes are small, the chi-squared test will give inaccurate results.

Solution

See Table 4.48.

Table 4.47 Synonymicity versus fixity

	Synonymous	Replacement
Polymorphisms	43	2
Fixed	17	7

Table 4.48 Table of calculation Fisher's exact test

	Synonymous	Replacement	Total
Polymorphisms	**43** − a	**2** − b	**45** − a + b
Fixed	**17** − c	**7** − d	**24** − c + d
Totals	**60** − a + c	**9** - b + d	**69** − n

$$p = \frac{\binom{a+b}{a}\binom{c+d}{c}}{\binom{n}{a+c}} = \frac{(a+b)! + (c+d)! + (a+c)! + (b+d)!}{n!a!b!c!d!}$$

The exact probability of this table, **p = 0.00605.**

Probability of all possible tables with the same marginal totals (60, 9, 45, and 24).

The p-value is the sum of all probabilities less than or equal to **0.00605.** Therefore, p-value = 0.000584 + 0.00605 + 0.000023 = **0.006657**.

45	0	45
15	9	24
60	9	69

p = 0.000023

42	3	45
18	6	24
60	9	69

p = 0.0337

43	2	45
17	7	24
60	9	69

p = 0.00605

44	1	45
16	8	24
60	9	69

p = 0.000584

41	4	45
19	5	24
60	9	69

p = 0.1117

Conclusion: At significance level 0.05, the null hypothesis of independence is rejected because the p-value 0.006657 < 0.05.

Case Study 2

Custer and Galli (2002) flew a light plane to follow great blue herons (Ardea Herodias) and great egrets (Casmerodius albus) from their resting site to their first feeding site at Peltier Lake, Minnesota and recorded the type of substrate each bird landed on (Table 4.49).

Suitable tool: Randomization test of independence.

Wrong tool: Chi-square test.

Table 4.49 Type of substrate at Peltier Lake

	Blue Heron	Great Egrets
Vegetation	15	8
Shoreline	20	5
Water	14	7
Structures	6	1

Reason: There are two nominal variables, the sample is small, and there are many categories.

Solution
A randomization test with 100,000 replicates yields $p = 0.54$, so there is no evidence that the two species of birds use the substrates in different proportions.
Estimation and Confidence intervals

Estimation [13]

Consider these real-life situations,

- Students estimate the scores they may get after discussing the answers with their peer group.
- Google maps give you an estimated time to reach your destination.
- While shopping, we try to estimate if the money in the wallet can afford the items in the cart.

This way estimation in the world of statistics is as common as it is in real life. In statistics, we estimate a population parameter (example, θ) with some facts known to us. The sample used for estimating is called estimator, the value of statistic drawn from this sample is called estimate and the entire procedure is called estimation.

Example, let a sample of 10 students be chosen to study the population. Then sample is the estimator, value of mean of sample = estimate and the procedure is called estimation (Fig. 4.26).

Point estimate: A point estimator of some population parameter θ is a single numerical value of a statistic. The table below shows the point estimators for different population parameters.

Interval estimate

When we are quite uncertain with locating estimators, we use interval estimates where there is some marginal error that is added on both sides of the point estimate. So, this becomes a range of plausible values for an unknown parameter θ. The endpoints of the intervals are referred to as the upper and lower confidence limits (Table 4.50)

Example 25 The electricity supply board for the Baadampur district surveys the average electricity consumption in the villages south of the district. For 86 villages, the mean and standard deviation of the number of units of consumption are 1950 and 250 units, respectively, for the first quarter of the year. Let's construct a 95% confidence interval estimate of the actual mean consumption. So that the Baadampur Electricity Board (BEB) can allocate a sufficient budget to provide electricity with some awareness of the possible unit consumption.

Solution

$\overline{X} = 1950$, $\sigma = 250$, $n = 86$, confidence interval $(1 - \alpha) = 95\%$, $Z_{\alpha/2}$ value is equal to $\pm$ 1.96.

Therefore,

$$\overline{X} \pm Z_{\alpha/2} \frac{\sigma}{\sqrt{n}} = 1950 \pm 1.96 \left(\frac{250}{\sqrt{86}} \right) = 1950 \pm 29.40.$$

The interval range is $[1896.16 \leq \mu \leq 2002.84]$.

At 95% confidence interval, the mean μ is likely to fall between 1896.16 and 2002.84 units. If the cost of supplying one unit of electricity is 4.25 rupees, the B.E.B must at least hold a budget of 8058.68 (4.25 * 1896.16) rupees.

Example 26 Settling claims at the earliest makes the company more reliable to customers. An insurance auditor would like to know the proportion of claims that the company is capable of settling within 3 months of receipt of the claim. To test this, a random sample of 150 claims settled was selected in the recent time period out of which 72 claims were settled within 3 months of time. As it is an insurance company with stringent claim policies, we set up a confidence interval of 99% to estimate the claim settlement ratio.

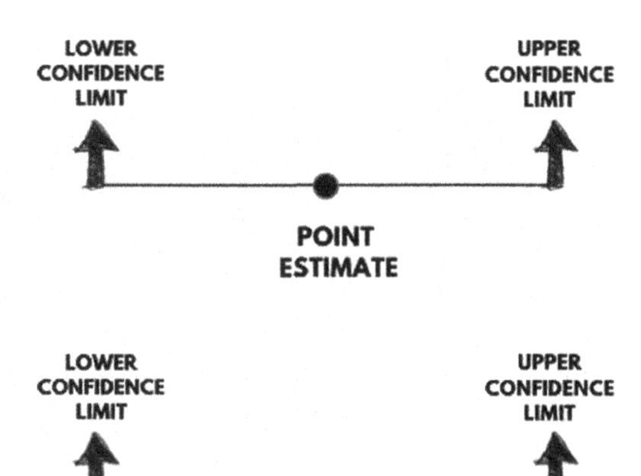

Fig. 4.26 Point and interval estimate

Table 4.50 Parameter, sample statistic and margin error for different tests

When	Parameter	Sample statistic	Margin error
σ is known. X is normal and $n \geq 30$	**Mean μ**	$\overline{X}$	$Z^*\left(\frac{\sigma}{\sqrt{n}}\right)$
σ is unknown. X is normal or $n \leq 30$	**Mean μ**	$\overline{X}$	$t^*_{n-1}\frac{s}{\sqrt{n}}$
$\widehat{np} \geq 10$ and $n(1-\hat{p}) \geq 10$	**Proportion** p	$\hat{p}$	$Z^*\sqrt{\frac{\widehat{p^*}(1-p)}{n}}$
$n_1\widehat{p_1}.n_1(1-\widehat{p_1})$ and $n_2\widehat{p_2}.n_2(1-\widehat{p_2})$ both are greater than 10	**Differences of proportions** $p_1 - p_2$	$\widehat{p_1} - \hat{p}_2$	$Z^*\sqrt{\frac{\widehat{p_1^*}\left(1-\widehat{p_1}\right)}{n_1} + \frac{\widehat{p_2^*}\left(1-\widehat{p_2}\right)}{n_2}}$
σ_1, σ_2 are known. And X_1, X_2 are normal and $n_1, n_2 \geq 30$	**Differences of mean** $\mu_1 - \mu_2$	$\widehat{X_1} - \overline{X}_2$	$Z^*\sqrt{\frac{\sigma_1^2}{n_1} + \frac{\sigma_2^2}{n_2}}$
σ_1, σ_2 are unknown. and $\sigma_1 \neq \sigma_2$ and X_1, X_2 are normal with $n_1, n_2 < 30$	**Differences of mean** $\mu_1 - \mu_2$	$\overline{X_1} - \overline{X}_2$	$t^*_{n_1+n_2-2}\sqrt{\frac{(n_1-1)s_1^2+(n_2-1)s_2^2}{n_1+n_2-2}}$
X_1, X_2 contain matched pairs, differences are normal with $n_d < 30$	**Differences of mean** $\mu_1 - \mu_2$	$\overline{X_1} - \overline{X}_2$	$t^*_{n-1}\frac{S_d}{\sqrt{n_d}}$

Note The sample mean in the table 4.50 is denoted by $\overline{\mathbf{X}}$

Solution

The sample proportion is given as $\frac{\text{Number of claims settled}}{\text{Total number of claims}} = \frac{72}{150} = 0.48$.

The $Z_{\alpha/2}$ value at 99% confidence interval is given as 2.576.

Therefore, the estimated interval will thus be

$$\overline{p} \pm Z_{\frac{\alpha}{2}}\sqrt{\frac{\overline{p}(1-\overline{p})}{n}} = 0.48 \pm 2.576\sqrt{\frac{0.48 * 0.52}{150}} = 0.48 \pm 0.10304$$

Therefore, the proportion of claims settled by the company within 3 months is likely to be between 37.6 and 58.3% As $0.37696 \le \overline{p} \le 0.58304$.

When to use which test? [14]

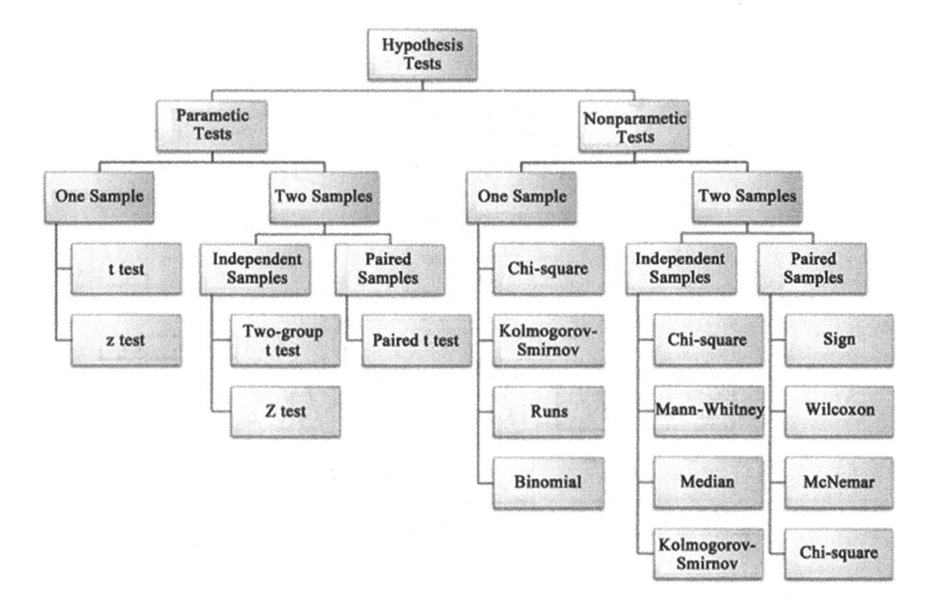

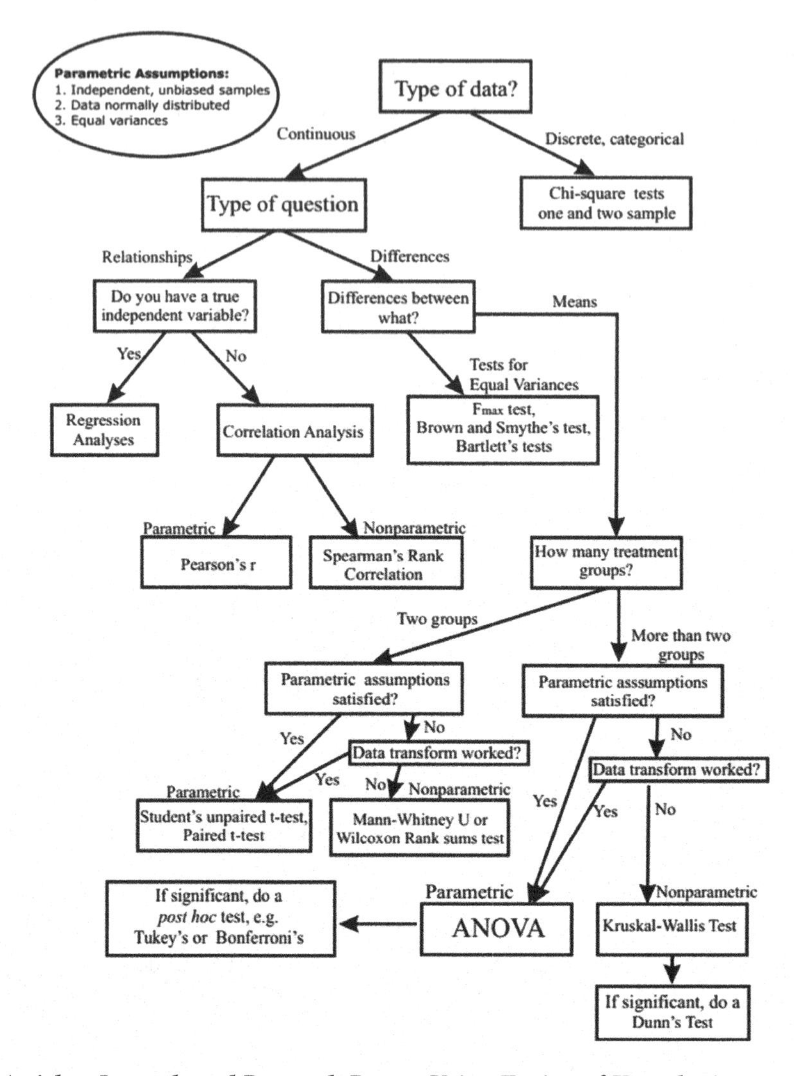

Articles, Journals and Research Papers Using Testing of Hypothesis

1. **Journal**: American Heart Association journal Title: Statistical Primer for Cardiovascular Research -Hypothesis Testing Authors: Roger B. Davis, Kenneth J. Mukamal URL: http://circ.ahajournals.org/content/114/10/1078.full
2. **Journal**: Journal of technology education Volume 7, Number 1, Fall 1995 Title: Collaborative Learning Enhances Critical Thinking In this study, the hypothesis testing procedure is used to compare two groups of undergraduate college students who underwent two different methods of instruction. Author: Anuradha A. Gokhale URL: http://scholar.lib.vt.edu/ejournals/JTE/jte-v7n1/ gokhale.jte-v7n1.html

3. **Why, how and When to use Post Hoc Tests**: https://statisticsbyjim.com/anova/post-hoc-tests-anova/
4. **ANOVA Practice Problems**: http://eagri.org/eagri50/STAM101/pdf/pract11.pdf
5. **Other Internet resources**: https://sigmazone.com/beyond_t_and_f_test/ https://www.scirp.org/journal/paperinformation.aspx?paperid=109618 https://online.datasciencedojo.com/blogs/type-i-and-type-ii-errors-smoke-detector-and-the-boy-who-cried-wolf.

Some Research Articles in Sum of Squares Using One-Way ANOVA

1. **Journal**: Journal of Vocational and Technical Education: Volume 14 Number 1 (1997) Title: Effects of Economic Disadvantaged Status and Secondary Vocational Education on Adolescent Work Experience and Postsecondary Aspirations. Authors: Jay W. Rojewski, The University of Georgia Illustrates the use of a 2x3 analysis of variance with strength-of-effect estimates computed for all significant results. URL: http://scholar.lib.vt.edu/ejournals/JVTE/v14n1/JVTE-4.html
2. **Electronic Journal of Human Sexuality**: Vol. 1, Sept. 15, 1998 Title: College Students' Perceptions of Women's Verbal and Nonverbal Consent for Sexual Intercourse. Authors: Jason J. Burrow, Roseann Hannon, and David Hall, University of the Pacific Illustrate the use of a 2x2 analysis of variance in which both factors consent to have sex (yes, no) and gender (male, female) was active factors. This analysis was performed twice, once of each of the 2 dependent variables. Tests of simple main effects were conducted to probe the significant interaction that was found in one of the analyses. URL: http://www.ejhs.org/volume1/burrow/burrow.htm
3. **Analysis of Variance—Components for Genetic Markers with Unphased Genotypes**: Published online 2016 Jul 13. Author: Division of Biostatistics, Institute for Health and Society, Medical College of Wisconsin, Milwaukee, WI, USA. Edited by: Steven J. Schrodi, Marshfield Clinic Research Foundation, USA. Reviewed by: Zhan Ye, Marshfield Clinic, USA; Charles M. Rowland, Quest Diagnostics, USA. https://www.ncbi.nlm.nih.gov/pmc/articles/PMC4942470/
4. **Studies in crop variation**: I. An examination of the yield of dressed grain from Broadbalk. Fisher, Rory A.. "Studies in crop variation. I. An examination of the yield of dressed grain from Broadbalk." ***The Journal of Agricultural Science 11: 107–135.*** https://www.semanticscholar.org/paper/Studies-in-crop-variation.-I.-An-examination-of-the-Fisher/d675c0558ffd08e96d08ba339620012e926c15c0
5. **ANOVA terminologies**: https://studylib.net/doc/5839473/anova-terminology

Practice Data Sets—Coding

Data set on air quality in Indian cities: We can understand the detailed process of retaining only the relevant information and also standardising the data using the standard normal distribution for analysis. https://github.com/naren951/Air-Quality-Index-Analysis/blob/master/AQI.ipynb

Data set from an electric car-sharing service company called Autolib: The project is about using hypothesis testing and investigating a claim

on blue car usage rates in Paris and Hauts-de-Seineconcept (French cities). https://github.com/JoanYego/Hypothesis-Testing-Autolib-Dataset/blob/master/Hypothesis_Testing_(Autolib_Project).ipynb
Market research for a box company from Tustin, CA: This research consists of responses from 5,000 Californians. Concepts from interval estimation and hypothesis testing are used to predict the preferred type of boxes that will be sold in the next few months. https://github.com/ghayward/hypothesis_testing_polling_data/blob/master/polling_question_python_answers.ipynb
Northwind database: To understand the regional and seasonal performances of Northwind traders Understanding the applications of t-test, Mann–Whitney U test, ANOVA and the central limit theorem (CLT). https://github.com/bonniema/HypothesisTesting-NorthwindEcommerce/blob/master/HypothesisTesting_Project.ipynb
Predictive modeling for car prices: Analysing data attributes and their categorisations. Feature selection and correlation. Regression and Prediction https://github.com/elena-petrova/predictive_modeling_price/blob/master/Predictive_Modeling_Price.ipynb
Practice data Sets: Hypothesis testing with lens correction data: https://sigmazone.com/lens-hyptest/

Other References

1. **Acceptance and rejection regions**: https://ani.stat.fsu.edu/~debdeep/p10.pdf
2. **One and two tailed tests**: https://www.sciencedirect.com/topics/mathematics/tailed-test
3. **p-value**: https://www.simplypsychology.org/p-value.html
4. **Real life problems translated into hypothesis language**: https://www.researchgate.net/figure/2-Three-Approaches-of-Hypothesis-Testing-16_tbl6_263846563
5. **Errors and types of errors**: https://bit.ly/2XFv6oh, https://online.datasciencedojo.com/blogs/type-i-type-ii-errors/: Casey Anthony, O.J. Simpson and other examples attempt to minimize the extent to which innocent individuals are falsely convicted
6. **Scandal on a medical device maker**: Adapted from European Journal of Epidemiology, April; 25(4): 223–224. Curbing type I and type II errors by Kenneth J. Rothman.
7. **Between and within variations**: https://bookdown.org/BaktiSiregar/data-science-for-beginners-part-2/11-ANOVA.html
8. **Post Hoc Tests**: https://bit.ly/3mxISBZ
9. **Tukey's simultaneous 95% CI's**: https://www.researchgate.net/figure/Tukey-Simultaneous-95-Confidence-Intervals-for-Taber-Abrasion-Resistance-Coating_fig25_303719917
10. **Binomial tests**: https://math.stackexchange.com/questions/2123873/is-the-maximum-of-a-probability-distribution-function-of-a-binomial-distribution
11. **Goodness of fit**: https://www.akira.ai/glossary/goodness-of-fit/
12. **Interpretation of Cochran's Q test**: https://www.scalestatistics.com/cochrans-q.html
13. **Comic on Estimation**: https://comicsidontunderstand.com/tag/frank-ernest/
14. **When to use which test?** https://www.scirp.org/journal/paperinformation.aspx?paperid=109618

Chapter 5
Correlation and Regression

WHAT

Concepts of correlation and regression of different types with examples and formulae.

WHY

- To measure the degree of relationship between the variables.
- These are vital concepts predominantly used in all areas of research and data analytics.

HOW

- Through various statistical formulae.
- Through scatter plots.

WHEN

- To check if there exist any relationship between the variables under study. If yes, then to study the nature of such relationship.
- Correlation and regression are the base for predictive analytics.

S. Prasad, *Elementary Statistical Methods*,
https://doi.org/10.1007/978-981-19-0596-4_5

CHAPTER 5

WITH MR.STAT

- Correlation examples and types.
- Calculating and interpreting correlation.
- Various types of correlation.
- Regression and its calculation.
- Types of regression.
- Differences in ANOVA and regression

WITH MISS TICS

- Case studies using correlation in Priory medical journals.
- Dealing with fuel efficiency data set using Python
- Inter- Commodity ETF correlations
- Different spurious correlations

5.1 Introduction: Correlation

The dictionary definition of the term correlation is: "*A mutual relationship or connection between two or more things*". A more formal definition says, "**Correlation is a statistical method used to assess a possible association between two or more variables**". In statistics, correlation is an indispensable tool that forms the basis for in-depth statistical analysis like forecasting, decision-making and simulation. Correlation deals with three main questions:

1. Is there a relationship between variables under study?—Correlation coefficient answers this (0 means no linear correlation, any other value denotes the presence of correlation)
2. What is the strength of this relationship? Value of correlation coefficient (value closer to $+1$ or -1 is strong and closer to 0 means lesser correlation)
3. What is the direction of this relationship? Given by the sign of correlation coefficient (positive or negative) (Fig. 5.1).

Let's understand the concept and calculation of this essential tool with various examples and data sets.

Examples:

- We always observe that seasonal change correlates with sales of a lot of products like our ice creams, umbrellas and woollen clothes.
- A correlation exists between watching violent content on television for long hours and the aggressive behaviour of adolescents.
- Psychologists agree that there is a high correlation between a toddler's hyperactiveness and the percentage of sugar in the food consumed.
- Econometrics deals with a lot of correlations between price and quantity, demand and supply of goods and services etc.

Fig. 5.1 Correlation between efforts and rewards [1]

Table 5.1 Types of variables used in correlation

Cause	Effect
Independent	Dependent
Predictor	Predicted
Explanatory	Response

Variables and types of variables:

Explanatory, response and lurking variable: Consider the ordered pair (x, y), where **x is called the explanatory variable and y is the response variable**. According to the linear model, changes in x have a corresponding change in y. In analytics, the variable that explains a scenario or which can be manipulated in experiments is called explanatory variables. The variable that depicts the effect of changes in the explanatory variable is called the response variable. Table 5.1 shows the different names of the cause-and-effect variables. It must, however, be noted that in correlation, we do not distinguish between" cause-and-effect" variables as correlation does not deal with "causation". This concept has been dealt with, in detail, a little later in the same chapter.

The Lurking variable: Lurking variable is responsible for changes in both **x and y**. Many times, such a variable causes a correlation between two related or unrelated variables. It can be defined as a variable that has an important effect but is not included among the predictor variables under consideration. Consider the examples below.

Example 1: A study may show a high positive correlation between the number of ice creams sold and the number of deaths from drowning. It is definitely wrong to conclude that eating ice cream leads to more deaths!

Analysis: The lurking variable, in this case, is temperature. Hotter days mean more ice creams and more people go swimming thus, this variable causes an increase in both variables under study.

Example 2: People who drink bottled water have larger babies than those who drink tap water.

Analysis: Here, the lurking variable is affluence, only rich people can afford to drink bottled water and their babies have larger birth weight too. Thus, it will be wrong to conclude that bottled water causes larger babies.

Example 3: Over the last decade, the population of a certain locality showed an increase.

If x denotes the number of people who pray in places of worship.

y: is the number of people who were caught speeding on roads. And **x** and **y** showed a high degree of correlation; it is wrong to conclude that praying caused people to speed on roads. The lurking variable here is the increase in population.

Concept of Covariance

Both correlation and covariance measure the relationship between two or more variables. But covariance is used to measure the joint variability of two sets. A positive covariance implies a direct relationship between the data sets and a negative covariance implies that the data sets are inversely proportional to each other. Also, the higher the covariance value, the higher the dependency of data. Covariance is modelled using the expectation function of the two data sets X and Y.

$$\text{Covariance (X, Y)} = \text{Cov(X, Y)} = \text{E[XY]} - \text{E[X] E[Y]}$$

Scatter Plot: A simple visual tool used to easily map the data points and identify if they are correlated. This form of representation is used as a quick guide for 2-dimensional data. Many scatter diagrams are illustrated in this chapter.

Types of Correlation

1. **Positive correlation**: A type of correlation where the variables move/vary in the same direction. For example, the National Institute of Ageing, USA, through several research studies, has demonstrated a strong positive correlation between social interaction with health and well-being among older adults. This means, the higher the social interaction, the greater will be the well-being of senior citizens (Fig. 5.2).

   Negative correlation: A type of correlation where the variables move in opposite directions, which means the variables are inversely proportional. Examples: High expenditure and fewer savings, number of classes attended and grades of a student and so on.
2. **Simple correlation**: When simple correlation is a measure used to determine the strength and the direction of the relationship between two variables, X and Y.

Example: X = amount of fertilizer (Kg) per acre of land.
Y = Yield of a crop (100 Kg).

Multiple correlation: When we consider more than two variables for our study. This involves quantifying the linear relationship between one dependent variable and several independent variables. That is, we try to study the effect of many variables on a single variable.

Example: X = amount of fertilizer (Kg) per acre of land.
Y = Yield of a crop (100 Kg).
Z = Rainfall (cm) in the area.
We analyse the effect of X and Z on Y (Fig. 5.3).

Partial correlation: When we consider more than two variables for our study but study only two at a time, keeping the others constant. These will highlight the relationships between each predictor and the response, after adjusting for other predictors. The main difference between simple and partial correlation is that, in simple correlation, we consider only two variables and completely ignore others. But in partial

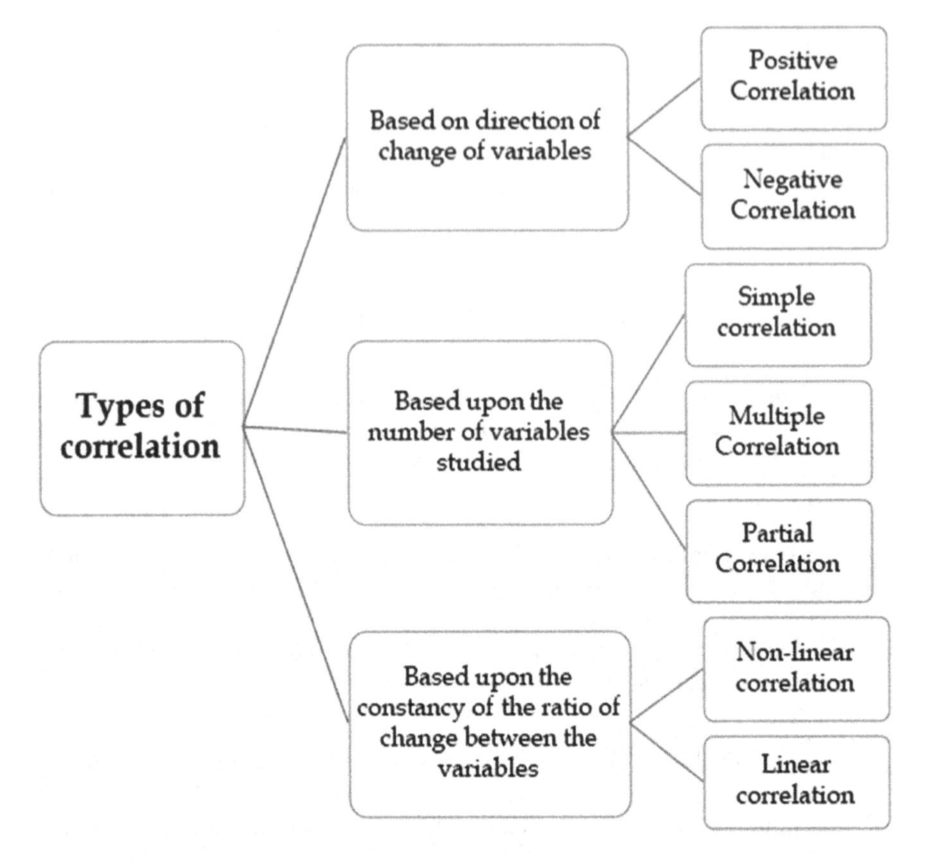

Fig. 5.2 Types of correlation based on different aspects

correlation, all variables which can influence our study are taken into account but only two of them are considered for the study.

Example: $X =$ amount of fertilizer (Kg) per acre of land.

$Y =$ Yield of a crop (100 Kg).

$Z =$ Rainfall (cm) in the area.

If we analyse the effect of X on Y, keeping Z constant, that is limiting our study to only those areas which are getting a particular amount of rainfall, then it is partial. Similarly, when we study the effect of Z on Y, keeping X constant that is, limiting our study to only a particular amount of fertilizer per acre, correlation is termed as partial.

(Detailed explanation given further down, in the same chapter)

3. **Linear Correlation**: A correlation is said to be linear if the ratio of change between variables is constant. When plotted on the scatter plot, the data points tend to lie close to a straight line. For example, usage and cost of electricity in a household.

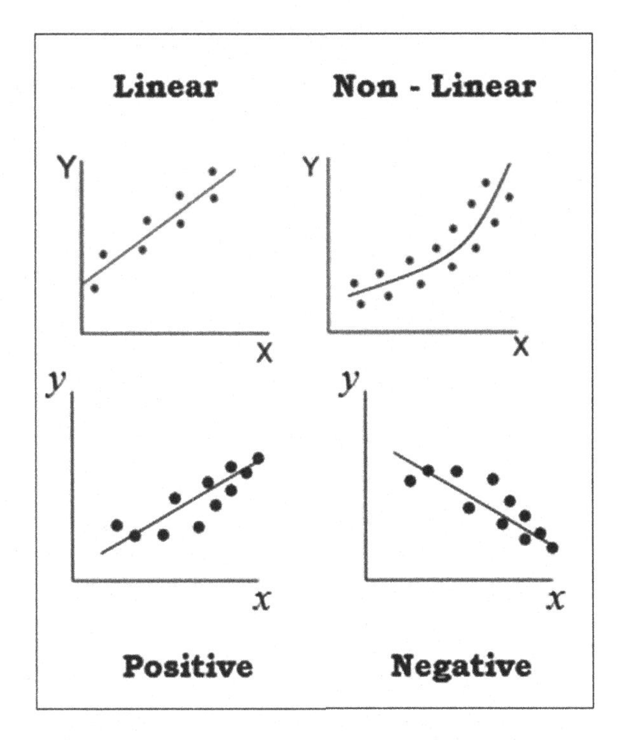

Fig. 5.3 Types of correlation charts

Nonlinear Correlation: A correlation is said to be nonlinear or curvilinear if the ratio of change is not constant. Here, all points on the scatter diagram tend to lie near a smooth curve.

Correlation also doesn't imply causation—The well-known warning.

This is the reason why we say that logic must always accompany correlation. When we find two variables with high correlation, they may not necessarily mean one causes the other. Even a high degree of correlation cannot be used to prove that one variable is the cause and the other is the effect. Correlation measures the strength of linear numerical relations and determination of cause and effect requires an in-depth knowledge of the subject and careful experimentation. Statisticians always say logic and correlation must go hand in hand. This is because even if numbers present a strong mathematical relationship between the variables, they must be logical and applicable to real-life scenarios.

A popular saying about correlation is "*The mind and body are perfectly correlated but not wealth and happiness*".

One might be tempted to conclude a cause and effect in a situation with a strong correlation but this could be due to sheer chance/coincidence/due to the presence of a lurking variable.

Consider the above chart which depicts crude oil imports from Norway and drivers killed in collisions with railway trains. The correlation between these two variables

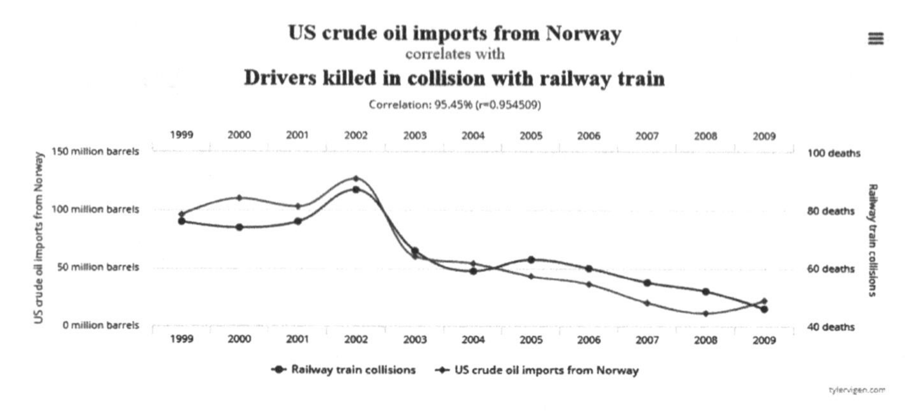

Fig. 5.4 Example of spurious correlation [2]

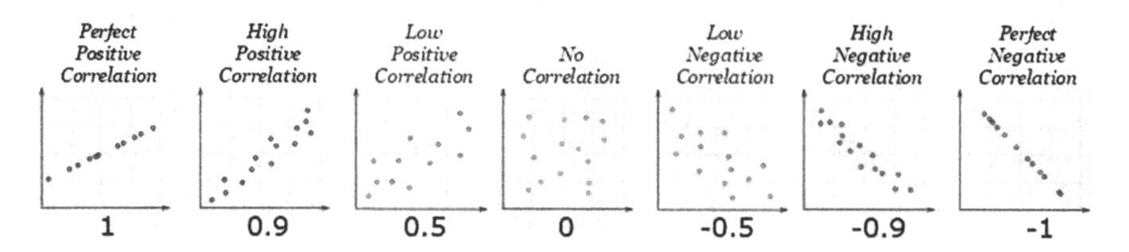

Fig. 5.5 Graphical representation of correlation coefficient

is very high, which is 95.45%. But in reality, analysing the correlation between these two variables makes no sense (Fig. 5.4).

The famous argument explaining such situations is the one in 1950 which says, "**Smoking causes Lung cancer**". There has been a sixfold increase in the rate of lung cancer in the preceding two decades. One school of thought was, that the increased rate could have been the result of more pollution due to industries and traffic that degrades air quality. But after intense research, doctors stated that smoking is one of the major risk factors for lung cancer.

Cigarettes and Cancer—An in-depth study of correlation and causation: [3]

Cause-and-effect relationships using Random experiments

Only random experiments can be used to infer cause-and-effect relationships. This is because only in such experiments, the effect of confounding variables will be the same for all groups. The differences in response variables are thus only due to the explanatory variables.

In the case of observational studies, there may be other differences between groups that may produce confounding variables and their effects will be difficult to separate from the effect of explanatory variables. Thus, the differences in the response variable cannot be attributed to differences in the explanatory variable.

Interpretation of correlation values (Fig. 5.5 and Table 5.2)

Table 5.2 Interpretation of coefficient of correlation

Range of values	Strength of linear relationship	Direction of linear relationship
−1 to −0.8	Strong	Negative
−0.8 to −0.6	Moderate	Negative
−0.6 to −0.3	Weak	Negative
−0.3 to 0.3	None	None
0.3 to 0.6	Weak	Positive
0.6 to 0.8	Moderate	Positive
0.8 to 1	Strong	Positive

Fig. 5.6 Karl Pearson

Karl Pearson's Correlation

A renowned British Mathematician and a Biostatistician whose contribution to the field of statistics is immense. Karl Pearson is known not just for correlation concepts, but also for chi-square and least square approximations. Karl Pearson's coefficient of correlation is also called the **product-moment correlation coefficient**. This is a simple measure of the linear relationship between two variables. It indicates the degree of correlation between these two variables. It can be calculated only if the characteristics under study are quantitative. Therefore, Karl Pearson's correlation coefficient between two random variables X and Y usually denoted by **r(X, Y)** is given by:

$$r(x, y) = \frac{\text{Covariance of } (X, Y)}{\sqrt{\text{Variance } (X) * \text{Variance}(Y)}}$$

$$= \frac{\text{Cov}(x, y)}{\sigma_x \sigma_y}$$

$$= \frac{n \sum XY - (\sum X)(\sum Y)}{\sqrt{n \sum X^2 - (\sum X)^2} * \sqrt{n \sum Y^2 - (\sum Y)^2}}$$

where, $1 \leq r(x, y) \leq 1$.

However, these are situations where Pearson's correlation has to be used with caution (Fig. 5.6).

1. **When data has outliers**: Outliers influence Pearson's correlations to a large extent. One option is to remove such outliers. In Spearman's correlation method, data is converted to ranks. Thus, outliers do not pose a problem and so, Spearman is more robust.
2. **When the data has highly skewed variables**: When computing correlation for skewed variables, a transformation like a log transformation is required which makes the underlying relationship between the two variables clearer.

Probable Error: The probable error (PE) of the coefficient of correlation indicates the extent to which its value depends on the condition of random sampling. If r is the calculated value of the correlation coefficient in a sample of n pairs of observations, then the probable error PE_r of the correlation coefficient r is given by

$$PE_r = 0.6745\left[\frac{1 - r^2}{\sqrt{n}}\right]$$

Probable error also helps in determining the reliability of the coefficient. Thus, with the help of PE_r we can determine the range within which the population coefficient of correlation is expected to fall using the following formula.

$\boldsymbol{\rho = r \pm PE_r}$ where ***ρ-rho*** represents the population coefficient of correlation. The PE defines the middle 50% of the normal distribution.

In Dr. Wheeler's words: "*No measurement should ever be interpreted as being more precise than plus or minus one Probable Error since your measurement will err by this amount or more at least half the time*".

Remarks

1. If the value of **r** is less than the probable error, there is **no evidence of correlation.**
2. If the value of **r** is more than six times the probable error, the coefficient is **significant**.
3. If the coefficient of correlation is 0.5 or more, the probable error is minimal. This way, correlation is generally considered to be significant.

Probable error as a measure for interpreting coefficient of correlation. The probable Error can be used only when the following three conditions are fulfilled:

- The data must be approximated to the bell-shaped curve which is a normal frequency curve.
- The Probable error computed from the statistical measure must have been taken from the sample.
- The sample items must be selected in an unbiased manner and must be independent of each other.

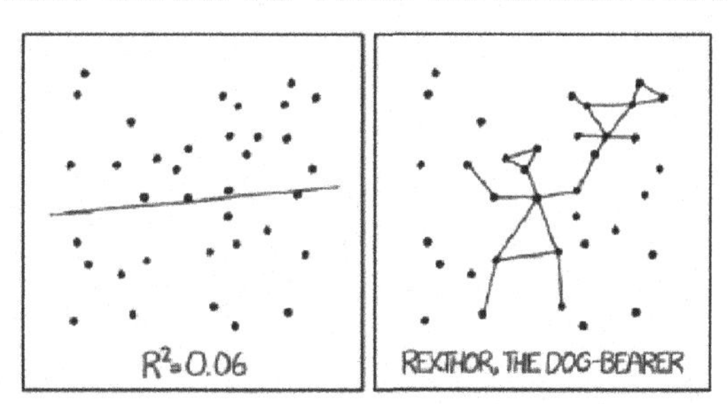

Fig. 5.7 Scatter plot with R squared value

Thus, the probable error is calculated to check the reliability of the value of the coefficient calculated from the random sampling.

Example 4: If **r** is +0.5 and $n = 5$, would you say that correlation is significant?

Solution:

Even though r = +0.5, its significance can be judged based on its P.E. only.

P.E. (r) = 0.6745 (1 − 0.25)/$\sqrt{5}$ = 0.2268 **and** 6 * PE = **1.3608**

Significance: "**r**" is more than PE. 0.5 > 0.2268, there is evidence of correlation.

The value 6 * PE = **1.3608**. In this case, r is not more than the 6 * PE value. 0.5 < 1.3608.

Hence it is not significant.

Conclusion: **r** is not significant even though its value is +0.5. This is so because ***n*** is very small.

It means that in general, the limits of the coefficient of correlation should be **r ± P.E**. In this example, 0.5 ± 0.2268 which is, 0.2732 and 0.7268**.**

These will be the limits of the correlation coefficient in the population.

Probable error as a measure for interpreting the coefficient of correlation should be used only when the number of pairs of observations is large. If ***n*** is small, probable error may give **misleading conclusions**, like the above example.

Example 5: **r** is +0.9 and $n = 100$, would you say that correlation is significant?

Solution:

P.E. = 0.6745 (1 − r^2)/$\sqrt{n}$ = **0.0128** and 6 * PE = **0.0768**.

Significance: "**r**" is more than PE. 0.9 > 0.0128, there **is evidence of correlation**.

The value 6 * PE = 0.0768. In this case r is more than the 6 * PE value, 0.9 > 0.0768 (Figs. 5.7 and 5.8).

Fig. 5.8 Charles Spearman

Hence there is a significant correlation.

It means that in the universe, the limits of the coefficient of correlation should be $r \pm$ P.E. Which is, 0 0.9 ± 0.0128 which is equal to 0.8872 and 0.9128 will be the limits of correlation coefficients in the population.

Coefficient of Determination (r^2)

This is the square of **r** and lies between 0 and 1. Most computer programs express this value as a percentage. The direction of correlation is lost through this expression as it is squared, but it gives the amount of variation as explained by **x**.

If $r = 0.5$, $r^2 = 0.25$ and the researcher may conclude that the explanatory variable explains 25% of the variation among values of the response variable.

Thus, this concept is extensively used by fund managers to check how much the fluctuations of the fund are explained by the moments in the benchmark index. If the r^2 value ranges from 0.85 to 1, that means the fund's performance is relatively in line with the benchmark index, which sometimes is a rare case. This value is intuitive as well as handy in the field of biology while experimenting on mutations to see quickly identify the gene that has a significant impact on the trait that is tested.

Spearman Rank Correlation.

This non-parametric version of the Pearson product-moment correlation was found out by **Charles Spearman**, an English psychologist**. Spearman rank correlation coefficient** quantifies the degree of linear association between the ranks of X and the ranks of Y.

Many psychological tests use Spearman's correlation to check the correlation of qualitative data such as honesty, intelligence, happiness. This rank correlation is also used when one or both of the variables consist of ranks when the data is continuous or in the ordinal or nominal scale of measurement. But this method cannot be used for grouped frequency distributions.

Formula:

$$\rho = 1 - \frac{6\sum {d_i}^2}{n(n^2 - 1)}$$

Table 5.3 Cost of advertisement and sales

Advertisement cost (in'000$)	39	65	62	90	82	75	25	98	36	78
Sales (in lakh$)	47	53	58	86	62	68	60	91	51	84

where d_i is the difference in paired ranks.

N is the number of cases.

When the ranks are tied, we use another formula:

$$P = \frac{\sum_i (x_i - \overline{x})(y_i - \overline{y})}{\sqrt{\sum_i (x_i - \overline{x})^2 \sum_i (y_i - \overline{y})^2}}$$

where i is the paired score.

C.F. $= \frac{m^3 - m}{12}$, where m is the number of times a rank is repeated.

In some cases, common ranks are assigned to repeated values. Suppose 80 is the highest score in the class and 2 students have secured the same marks. So, the average of first and second rank which is **1.5**{(1 + 2)/2} is assigned to both the students.

In case, three students have scored 80, the middle of ranks 1, 2 and 3 which is 2 is assigned to all the 3 students. The next best score gets ranked 4 and not 3.

Since repeated ranks are used, we add a small component called **correction factor CF** in the numerator. C.F $= \frac{m^3 - m}{12}$, Where m is the number of times a rank is repeated.

Karl Pearson's versus Spearman's Correlation coefficient:

1. When the assumptions of Pearson's constant variance and linearity are not met, the researcher can try Spearman's correlation method.
2. If there are more than 100 data points, and the data is linear, then Pearson's method will be very similar to Spearman's. But computations become tedious in Spearman's method.
3. If the data has some nonlinear components and linear regression may not be appropriate, then the researcher can try to straighten out the data into a linear form by applying a transformation, maybe a log transform. If it still is not linear, then Spearman may be appropriate.
4. Many times, researchers prefer Pearson's correlation because it enables direct comparability of findings and it can be used for regression too. Also, there is not much difference between Pearson and Spearman correlation coefficients in many cases.

Example 6: Calculate the coefficient of correlation from the advertisement cost and sales as per the data given below: [4] (Table 5.3).

Suitable tool: Karl Pearson's coefficient of correlation.

Reason: The variables involved are in a ratio scale, so Pearson's correlation is a suitable choice.

Table 5.4 Calculation of correlation for the cost of advertisement and sales

Advertisement (X)	Sales (Y)	X^2	Y^2	XY
39	47	1521	2209	1833
65	53	4225	2809	3445
62	58	3844	3364	3596
90	86	8100	7396	7740
82	62	6724	3844	5084
75	68	5625	4624	5100
25	60	625	3600	1500
98	91	9604	8281	8918
36	51	1296	2601	1836
78	84	6084	7056	6552
650	660	47,648	45,784	45,604

Calculation (Table 5.4):

$$r(X, Y) = \frac{(10 * 45604) - (650 * 660)}{\sqrt{10 * 47648 - 650^2} * \sqrt{10 * 45784 - 660^2}}$$

$$r(X, Y) = \frac{456040 - 429000}{\sqrt{476480 - 422500} - \sqrt{457840 - 435600}} = \mathbf{0.78041}$$

Conclusion: **r = 0.78041**. Thus, the correlation is quite high and it is also positive. This means the cost of the advertisement has a positive impact on sales.

Example 7: An economist is studying the job market in the Denver area and its neighbourhood. A sample of six Denver neighbourhoods gave the following information (units in 100 s of jobs) [5].

X: Total number of jobs in a given neighbourhood.

Y: Number of entry-level jobs in the same neighbourhood (Table 5.5).

Suitable tool: Karl Pearson's coefficient of correlation.

Reason: The variables involved are in ratio scale, so Pearson's correlation is the suitable choice.

Table 5.5 Number of entry-level jobs in a locality

Number of jobs (X)	16	33	50	28	50	25
Number of entry-level jobs (Y)	2	3	6	5	9	3

Conclusion: **r = 0.860**. There exists a high degree of correlation between the total number of jobs in a given neighbourhood and the number of entry-level jobs in the same neighbourhood.

Example 8: Calculate coefficient of correlation between age and successful candidates in the examination [6](Table 5.6).

Age of candidates	Candidates appeared	Successful candidates
13–14	200	124
14–15	300	180
15–16	100	65
16–17	50	34
17–18	150	99
18–19	400	252
19–20	250	145
20–21	150	81
21–22	25	12
22–23	75	33

Suitable tool: Karl Pearson's coefficient of correlation.

Reason: The variables involved are in ratio scale, so Pearson's correlation is the suitable choice.

Calculations:

(i) The class intervals of age have to be represented by their mid-values which is X.
(ii) No of successful candidates have to be calculated on a common base which is, percentage or per thousand or any other such value. For example, in the class interval 13–14, (Y = 124/200) * 100 = 62 and so on.

$$r(X, Y) = \frac{(10 * 10417) - (180 * 588)}{\sqrt{10 * 3322.5 - 180^2} * \sqrt{10 * 35138 - 588^2}}$$

$$r(X, Y) = \frac{104170 - 105840}{\sqrt{33225 - 32400} * \sqrt{351380 - 345744}}$$

$$r(X, Y) = \frac{104170 - 105840}{\sqrt{825} * \sqrt{5636}} = \mathbf{-0.774468}$$

Conclusion:r = −0.7747. There exists a high degree of correlation between age and the success of candidates according to the given data.

Example 9: Consider the following ranks of top 8 final year Diploma students in essay writing and Calligraphy. Based on ranks, test the correlation of scores in these subjects that are closely related to each other (Tables 5.7 and 5.8).

Table 5.6 Calculating correlation between age and success of candidates

Age of candidates	Candidates appeared	Successful candidates	Y	X	XY	X^2	Y^2
13–14	200	124	62	13.5	837	182.25	3844
14–15	300	180	60	14.5	870	210.25	3600
15–16	100	65	65	15.5	1007.5	240.25	4225
16–17	50	34	68	16.5	1122	272.25	4624
17–18	150	99	66	17.5	1155	306.25	4356
18–19	400	252	63	18.5	1165.5	342.25	3969
19–20	250	145	58	19.5	1131	380.25	3364
20–21	150	81	54	20.5	1107	420.25	2916
21–22	25	12	48	21.5	1032	462.25	2304
22–23	75	33	44	22.5	990	506.25	1936
Totals			588	180	10,417	3322.5	35,138

Table 5.7 Table displaying scores of students in essay writing and calligraphy

Essay writing	8	4	6	1	6	6	3	2
Calligraphy	8	2.5	5	4	2.5	1	6	7

Table 5.8 Calculating rank correlation with correction factor for scores of students

Essay writing	Calligraphy	d_1	d_i^2
8	8	0	0
4	2.5	−1.5	2.25
6	5	−1	1
1	4	3	9
6	2.5	−3.5	12.25
6	1	−5	25
3	6	3	9
2	7	5	25
Total			83.5

Rank 6 in statistics is repeated 3 times, therefore $m = 3$.

Correction Factor

$$C.F_3 = \frac{m^3 - m}{12} = \frac{27 - 3}{12} = 2$$

Rank 2 and 3 is shared between two students. Therefore 2.5 is seen twice. The correlation factor for 2 is,

$$C.F_2 = \frac{m^3 - m}{12} = \frac{8-2}{12} = 0.5$$

$$\text{Spearman's correlation} = 1 - \frac{6(\sum d^2 + C.F)}{n^3 - n}$$

$$1 - \frac{6 * (83.5 + 2 + 0.5)}{512 - 8} = \mathbf{0.0238}$$

Conclusion: The correlation between the ranks of students in calligraphy and essay writing is negative and the association is very poor. Creative talent in form literature and art are different even in reality. Therefore, the correlation values are logical to an extent.

Case Study 1

Consider a study on the volume of Facebook and Twitter shares that a web page generates, done by "**Search metrics**" a search and social analytics company. The study reveals that this volume is closely correlated to how high it ranks in Google searches and finds that top brand websites appear to have a natural advantage for ranking highly in searches. Search metrics analysed search results from Google for 10,000 popular keywords and 300,000 websites to pick out the issues that correlate with a high Google ranking.

Suitable tool: Spearman's rank correlation coefficient.

Reason: The correlations were calculated using Spearman's rank correlation coefficient. As the data can be ranked rather than measured, Spearman's rho is a better choice in this case.

Case Study 2

Consider this study about the strength of the link between the price of a convenience item (a 50 cl bottle of water) and distance from the Contemporary Art Museum in El Raval, Barcelona (Table 5.9).

Suitable tool: Spearman's rank correlation coefficient.

Reason: Spearman's rho is a suitable choice here. The distance from CAM, as well as the price of the bottle, are quantitative variables. They can easily be ranked. Spearman's method is more suitable here as we need to check for the distance from a particular place which is a measure to be ranked (closer to, farther than and so on).

Solution:

$$\rho = 1 - \frac{6(\sum d^2 + C.F)}{n^3 - n} = 1 - \frac{6(261.5 + 0.5)}{10^3 - 10} = \mathbf{-0.5879}$$

Table 5.9 Variables explaining the strength of the link

Convenience store	Distance from CAM (m)	Price of 50 cl bottle (€)
1	50	1.80
2	175	1.20
3	270	2.00
4	375	1.00
5	425	1.00
6	580	1.20
7	710	0.80
8	790	0.60
9	890	1.00
10	980	1.15

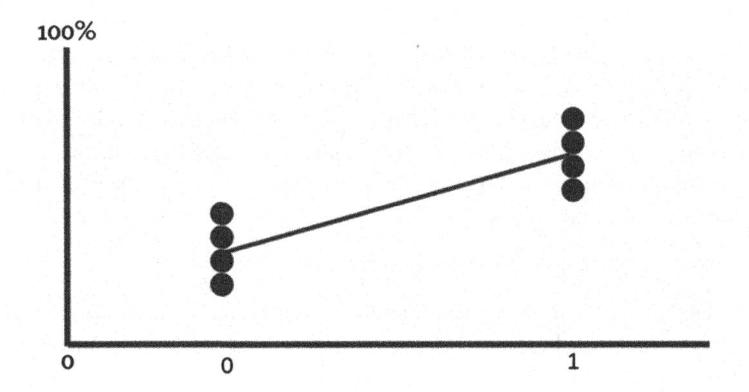

Fig. 5.9 Biserial correlation

Hence, Spearman's rank correlation coefficient is **−0.5879**. There is a moderate degree of negative correlation between distance from CAM (m) and price of 50 cl bottle (€) water bottle.

Case Study 3

Consider two teachers evaluating a student's answer sheet may be very lenient, some very strict. Consider a student's score as 55 from one examiner and 60 from the second teacher. His scores may be different, but when we compare his ranking among other students, it might almost be the same as the teachers would have maintained the same strictness/ leniency for all students (Fig. 5.9).

Suitable Tool: Spearman's rank correlation.

Reason: For such cases, rank correlation is more suitable as the discrepancies and differences in measurements will not affect the ranking. Product-moment correlation is not a suitable measure in this case.

Other Methods of Correlation

I. Biserial and point biserial correlation

When one variable can be measured in interval or ratio scale and the other can be measured and classified into two categories only, then **biserial correlation** has to be used. It is important to note that the second variable is continuous and normal.

For example:

1. Mental ability (measured using psychological tests) and result of an exam (success/failure) of 50 school children.
2. Amount of investment made with respect to the nature of the industry, like small scale/ large scale business.

Gravette and Wallanau (2009), in *Statistics for the Behavioral Sciences*, give some examples of dichotomous variables:

- College graduate versus not a college graduate.
- Firstborn child versus later-born child.
- Success versus failure on a particular task.

If the division is artificial, use a **coefficient of biserial correlation**. If it is natural, use the coefficient of **point biserial coefficient.**

Caution 1: Before applying biserial correlation, it must be tested for continuity and normal distribution of the dichotomous variable. This gives a better estimate when the split is around the middle, i.e., 0.5. The sample size should be fairly large too.

Caution 2: When the distribution is skewed and the split is not in the middle, this type of correlation is not suitable.

Caution 3: This correlation is not limited to the range -1 to $+1$. Therefore, it cannot be used for comparison with other correlation coefficients and also it is not suitable for regression analysis.

Caution 4: Check whether the dichotomy is natural or artificial. Natural means we do not apply an artificial point of division just for our convenience. For example, agriculturist/non-agriculturist, living in a city/not living in a city and so on.

In artificial dichotomy, the division is made because it is convenient and if the cut point is changed, many might move from one category to another. For example, if the pass mark is changed to 40 from say, 35, many might get shifted to the other category (Table 5.10).

II. **Tetrachoric correlation and phi coefficient** Consider a situation where both variables are dichotomous. It can be either a genuine dichotomy or made dichotomous for convenience sake. Then, biserial and point biserial cannot be used. We need to use tetrachoric correlation or phi coefficient. Then the Pearson's correlation calculated is called the **phi coefficient** φ (Fig. 5.10).

Table 5.10 Differences between biserial correlation and point biserial correlation

Bi serial correlation	Point bi serial correlation
Used when one variable is continuous and another one is artificially reduced to two categories	Used when one variable is continuous and another one is naturally dichotomous
Assumptions of normality, continuity, large N, split near median	No assumptions
Cannot be used in regression	Can be used in regression
Cannot be checked against "r"	Can be checked against "r"
No limits	Limits are between +1 and −1
No standard error of estimate exists	Standard error can be determined exactly and its significance can be tested against null hypothesis
Cannot be used when dichotomy is natural	Can be used even if there is a doubt as to whether dichotomy is artificial or natural
$r_b = [(Y_1 - Y_0) * (pq/Y)]/\sigma_{y'}$ Where: • Y_0 = mean score for data pairs for $x = 0$, • Y_1 = mean score for data pairs $x = 1$, • q = proportion of data pairs for $x = 0$, p = proportion of data pairs for $x = 1$, σ_y = population standard deviation Y is the height of the standard normal distribution at z, where $P(z' < z) = q$ and $P(z' > z) = p$	$r_{pb} = \left(\frac{Y_1 - Y_0}{S_Y}\right)\sqrt{\frac{np_0(1-p_0)}{n-1}}$ where $S_Y = \sqrt{\frac{\sum_{k-1}^{n}(Y_k - \overline{Y})^2}{n-1}}$, $\overline{Y} = \sum_{k=1}^{n} Y_k$, $p_1 = \frac{\sum_{k=1}^{n} X_k}{n}$, $p_0 = 1 - p_1$

Tetrachoric correlation is used when we have an artificial dichotomy and neither of them is in terms of continuous measures. Phi coefficient is used when we have a natural dichotomy and again, neither of them is in terms of continuous measures. It can be used when there is an artificial dichotomy too.

The tetrachoric correlation estimates are similar to what we would have obtained with Karl Pearson's coefficient for a 2 by 2 table of observed variables from a bivariate normal distribution. Therefore, φ is just Person's r on observed values. When we need to test the significance of φ. Take a look at Fig. 5.8.

Contingency tables: They cover all contingencies for the combinations of the two variables. They are also known as two-way tables as the categories of the two variables are used to create the table. *Generally, explanatory variables are written in rows and response variables in columns.*

Calculation: Tetrachoric correlation is obtained by the function given below, where *a*, *b*, *c* and *d* are the values in the cell.

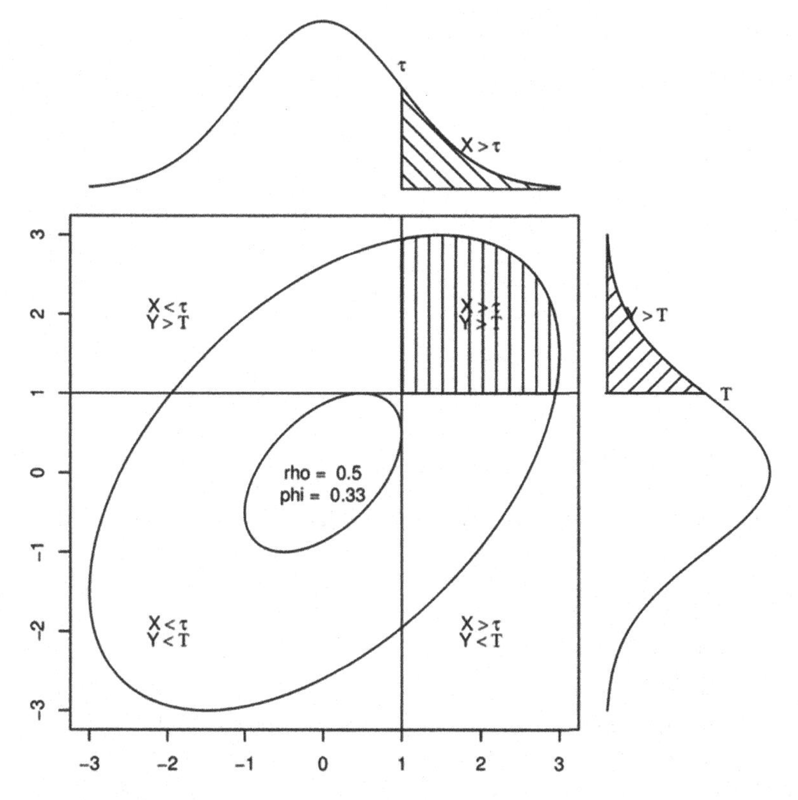

Fig. 5.10 Diagrammatic representation of tetrachoric correlation

$$R = \cos\theta$$

$$r_{tet} = \cos\left[\frac{180^\circ}{1+\sqrt{\frac{ad}{bc}}}\right]$$

$$\phi = \frac{AD - BC}{\sqrt{(A+B)(C+D)(B+D)(A+C)}}$$

Consider the following cases:

Case I: **X** is the variable denoting per capita income.

Y is the variable denoting the number of medical doctors per 10,000 residents.

Suitable tool: Pearson's product-moment correlation.

Reason: Here, we need to use Pearson product-moment correlation as both variables, *X*: income and *Y*: number of medical doctors, are at the ratio level as differences between measurement values are meaningful and theoretically can be zero. In Karl

Table 5.11 People and their best time to work

Peak time	0	0	0	0	0	1	1	0	1
Performance	65	80	55	70	55	40	60	60	60

Table 5.12 Result in a math test, an example of biserial correlation

Scores on a test of achievement	Result in the math test	
	Pass	Fail
100–200	0	4
200–300	2	5
300–400	10	15
400–500	24	4
500–600	30	0

Pearson's correlation, the variables must be interval or ratio scale and thus non-dichotomous.

Case II: **X**: Age of a student.

Y: Pass (1) or Fail (0) in an exam.

One of the variables, namely **Y** is dichotomous and the other is at the interval or ratio level, then the point biserial correlation coefficient is a better tool.

Case Study 4

*S*ome people think they do their best work early in the morning, whereas some claim they are at their best at night. Thus, we have a dichotomized situation here with morning people being marked as 0 and night people as 1. The independent estimates of the quality of work they do on a specified day is rated on a 100-point scale and is given below (Table 5.11).

Suitable tool: Point biserial correlation.

Reason: Here, the dichotomy is genuine/natural, as we can clearly distinguish between morning and evening. Thus, point biserial correlation has to be used here.

Case Study 5

The following table gives the distribution of scores on an achievement test. The students are divided as A pass/fail in a math test (Table 5.12).

Suitable tool: Biserial correlation.

Reason: In this case, pass/ fail is an artificial dichotomy as they can be changed according to our choice. The other variable is continuous and therefore, the suitable tool is a biserial correlation.

Table 5.13 Results in a math test, example of point biserial correlation

Scores on a test of achievement	Result in the math test	
	1-right response	0-wrong response
100–200	0	4
200–300	2	5
300–400	10	15
400–500	24	4
500–600	30	0

Solution:

$$r_{bis} = \frac{M_p - M_q}{\sigma_1} \text{x} \frac{pq}{y} = \frac{271.21 - 382.14}{110.27} \text{x} \frac{(0.7021 \text{x} 0.2978)}{0.3466} = \mathbf{-0.6068}$$

Case Study 6

The following table gives the distribution of scores on an achievement test. The students are divided as 0 if they gave the wrong answer and 1 if they gave the suitable answer, in a math test (Table 5.13).

Suitable tool: Point biserial correlation.

Reason: Observe that the data is the same. The classification of results is however natural as the answers can be suitable or wrong and there is no artificial dichotomy here. Thus, the point biserial coefficient is the best suited in this case.

Solution:

$$r_{p,bis} = \frac{M_p - M_q}{\sigma_1} \sqrt{pq} = \frac{271.21 - 382.14}{110.27} \text{x} \sqrt{0.721 \text{x} 0.2978} = -0.46$$

Therefore, the coefficient of point biserial correlation is **−0.46.**

Case Study 7

Variable **X** denotes intelligence which is further classified as intelligent and not intelligent.

Variable **Y** denotes emotional stability classified as stable/ unstable.

Table 5.14 Happiness based and exam results

Home situation (X)	Success in the exam (Y)	
	Success	Failure
Happy	45	25
Not happy	15	25

Table 5.15 Investment and financial success

Investment (X)	Financial success (Y)	
	Success	Failure
1	45	25
0	15	25

Note If a test is scored as pass/fail, like/dislike, agree/disagree, yes/no and if no intermediate responses are allowed, we can use phi correlation

Suitable tool: The division/dichotomy is artificial so the appropriate tool is tetrachoric correlation.

Case Study 8

Variable **X** denotes success in an exam classified as pass and fail. **Y** denotes the situation at home, happy and not happy (Table 5.14).

Suitable tool: Tetrachoric correlation.

Reason: *T*he division, as we observe, is an artificial dichotomy made according to our convenience, thus tetrachoric correlation has to be used here.

Solution:

$$r_1 = \cos\left(\frac{180\text{degrees of x } \sqrt{BC}}{\sqrt{AD} + \sqrt{BC}}\right) = 0.6085$$

Hence the coefficient of tetrachoric correlation is 0.6085.

Case Study 9

Consider the following X and Y dichotomous variables,

X: Nationality of a person classified as 1—Indian and 2—Others.

Y: Employment of a person classified as 1—Employed and 2—Unemployed.

Suitable tool: Phi correlation.

Reason: This is a true dichotomy as we can easily distinguish between Indians/non-Indians and employed/ unemployed persons (Table 5.15).

Case Study 10

Let **X**: Invested in mutual funds classified as 0—Not invested and 1—Invested, **Y**: Financial success.

Suitable tool: Phi correlation.

Reason: Observe that the numbers 1 and 0 represent genuine dichotomy. Success/failure are of course, dependent on the measuring instruments and phi correlation is a better measure.

Solution:

$$\phi = \frac{\mathrm{AD} - \mathrm{BC}}{\sqrt{(\mathrm{A}+\mathrm{B})(\mathrm{C}+\mathrm{D})(\mathrm{B}+\mathrm{D})(\mathrm{A}+\mathrm{C})}} = \frac{1125 - 375}{\sqrt{70 * 40 * 50 * 60}} = 0.259$$

Hence the coefficient of phi correlation is **0.259**.

Note 1: The phi correlation coefficient has the same relationship with tetrachoric as point biserial has with biserial coefficient.

Note 2: The relationship between chi-square and phi can be written as $\chi 2 = N\ \phi 2$

Note 3: When in doubt regarding the exact nature of the dichotomized variables, it is safer to use ϕ.

Some misconceptions about correlation coefficient are:

1. If the correlation between the scores on two variables is very high, then the two means must be very similar.
2. A correlation of 0.60 indicates twice the relationship strength as compared to a correlation of 0.30.
3. If a single outlier is removed from a very large group, the value of r cannot change very much.
4. An r of -0.90 signifies a low relationship.
5. If the correlation between two variables is equal to +0.50 for a subgroup of men, and if the correlation between these same two variables is +0.50 for a subgroup of women, then the combined correlation between these two variables will also be +0.50.
6. A linear relationship between two variables exists only if all the dots in a scatter diagram fall on a straight line.
7. If the researcher's data corresponds to two qualitative variables, it's impossible to compute a correlation coefficient.

Table 5.16 Salary and self-esteem

Subject number	Self esteem	Salary (Rs)
1	3	50,000
2	4	20,000
3	3	10,000
4	5	75,000
5	4	60,000
6	2	30,000
7	4	40,000
8	6	80,000
9	5	50,000
10	4	20,000

Some commonly committed mistakes while using correlation:

The assumption is that a correlation between pieces of data is proof of a cause-and-effect relationship.

Example 10: Consider the following data which gives salary in rupees and self-esteem measured on an appropriate psychological scale (Table 5.16).

Analysis:

In the above example, it will be wrong to conclude that self-esteem is affected by salary. Or the salary obtained is dependent on self-esteem. Though it is human tendency and judgement to assume causation, it is wrong in this case. The correlation coefficient, by itself, cannot be used to arrive at such a conclusion.

(i) If we want to prove, for some research, that self-esteem is indeed dependent on salary, we might deliberately choose such cases and try to conclude them. This is wrong as we need to use the data to arrive at conclusions and not the other way round.
(ii) Using simple correlation techniques in studies where partial correlation or multiple correlation is needed to obtain a clear picture of the way the variables are operating.

If we find a relationship between self-esteem and salary, it is an indication of

(a) A relationship between self-esteem and salary.
(b) Strength of the relationship.
(c) The direction of the relationship, positive or negative.

We may need to study more variables in order to get a clear picture of the relationship. These could be the level of education, family relationships and workplace equations and so on. A simple correlation would ignore these factors making the study incomplete. A deeper knowledge would lead to better analysis.

Fig. 5.11 Multiple correlation

Partial and Multiple Correlations.

In analytics, there are many variables involved and it is required to study these simultaneously and to test the relationship between them. In such a study, checking for relationships or differences between multiple variables is called **multivariate analysis**, and it is very important to check that this relationship is not spurious and there is a genuine relationship between the variables involved.

Age, for instance, is a factor that influences the relationship between many variables (Fig. 5.11).

For example, age influences height and weight, reading and writing skills, conservative approach to situations, income and so on. If the range of these measurements is high, these variables will surely exhibit correlation because of the common factors influencing both. Sometimes if the common factors are eliminated, these are generally eliminated by one of the following methods:

Fig. 5.12 Partial correlation

1. Eliminate the influences of these external variables through experimental control. This can be done by choosing only those units which possess a given characteristic (Fig. 5.12).

This is not a perfect method. For example, it will drastically reduce the sample size for the study. It involves meticulous identification of factors and this cannot be guaranteed in analytics. Also, it may not be possible to generalize this result to other groups.

2. Eliminate the influence of such variables statistically. A good design and technique involving statistics will help in controlling the variability of external factors. Such a technique is called the **partial correlation technique**.

Multiple correlation measures the combined influence of a group of variables on another variable, whereas partial correlation studies the relation between two variables when the effect of a third variable is removed.

Multivariate distribution is the combined distribution of three or more variables. It is usually observed that along with the independent and dependent variables, there is another variable, the third variable, which can be,

1. The **lurking variable** explains the relationship between **X** and **Y**.
2. The **Intervening variable is** intermediate between **X** and **Y**. Such variables are also called explanatory variables and are both the product of the independent variable as well as the cause of the dependent variable.
3. The **variable acting as a moderator,** in the sense that a direct relationship is observed for some samples but not for others. This can occur due to design or by chance.

There may be more than one independent variable determining the dependent variable. Almost all situations in analytics have a lot of variables affecting an event. For example, job satisfaction may be influenced by the nature of the job, interest in the specific project assigned, peer group, pay, etc. In such cases, it is not easy to determine which set of variables are independent variables. They have to be differentiated from intermediate or explanatory variables using logic and intuition.

Example 11: Let **X**: Course taken up by a student.

Y: Satisfaction obtained after studying the course.

Z: Interest in the subjects offered.

Here, Z is the intervening variable. It cannot be spurious as it is related to both the other variables, yet this does not fully explain the Y and it cannot be fully explained by X too.

Example 12: Let **X**: How much of the research work my job contains.

Y: Job satisfaction.

Z: Past job experience of the person.

Z is a moderator variable as this will influence both. Employees with more experience will be capable of handling multiple operations diligently than freshers in the company.

Multiple Correlation

With a single dependent variable and many independent variables, how well can you predict a variable using the linear relationship? This is answered by multiple correlations. R^2 represents the proportion of the total variance in the dependent variable that can be accounted for by the independent variables.

Coefficient of multiple correlation

The multiple correlation coefficient in terms of the total correlation coefficients between the pairs of variables is given by:

$$R_{1.23}{}^2 = \frac{r_{12}{}^2 + r^2{}_{13} - 2r_{12}r_{13}r_{23}}{1 - r^2{}_{23}}$$

$R_{1.23}$ denotes the multiple correlation coefficient of X_1 on X_2 and X_3. In other words, $R_{1.23}$ is the correlation coefficient between X_1 and its estimated value as given by the plane of regression of X_1 on X_2 and X_3. Other two correlations are similarly defined such as,

$$R_{2.23}{}^2 = \frac{r_{21}{}^2 + r_{23}{}^2 - 2r_{21}r_{23}r_{13}}{1 - r_{13}{}^2}$$

$$R_{3.23}{}^2 = \frac{r_{31}{}^2 + r_{32}{}^2 - 2r_{31}r_{32}r_{12}}{1 - r_{12}{}^2}$$

Partial correlation:

Partial correlation is a method used to describe the relationship between two variables when the effects of another variable, or several other variables, are removed. The independent, dependent and third variable are continuous and the assumptions of application are that of normality and linearity. It varies between −1 and +1 and its calculation is based on the simple correlation coefficient. In simple correlation, no factor is held constant, so it is also known as a zero-order coefficient. The partial correlation, studied between two variables by keeping the third variable constant, is called a **first-order coefficient**, as one variable is kept constant. Similarly, we can define a second-order coefficient and so on.

Examples:

1. An analyst wants to measure the correlation between exam anxiety and the past academic achievement of students keeping their IQ scores as constant.

2. For a medical analyst who wants to check the correlation of sugar levels and blood pressure levels in the patient keeping his heart rate to be a constant value (when the patient is relaxed).

When to use: when an analyst suspects that the relationship between 2 variables is influenced by other variables. It can be used in those cases where the phenomena under consideration has multiple factors influencing it and where it is possible to control the variables and the effect of each variable can be studied separately.

The partial correlation coefficient is said to be adjusted or corrected for the influence of the different covariates. The partial correlation of A and B adjusted for C is given by:

$$r_{ABC} = \frac{r_{AB} - r_{AC}r_{BC}}{\sqrt{(1 - r^2{}_{AC})(1 - r_{BC}{}^2)}}$$

The other partial correlations can be written similarly.
$r_{BCA} = \frac{r_{BC} - r_{BA}r_{CA}}{\sqrt{(1 - r^2{}_{BA})(1 - r_{CA}{}^2)}}$ and

$$r_{CAB} = \frac{r_{CA} - r_{AB}r_{CB}}{\sqrt{(1 - r^2{}_{CB})(1 - r_{AB}{}^2)}}$$

Caution:

1. The calculation of the partial correlation coefficient is based on the simple correlation coefficient which assumes a linear relationship. Generally, this assumption is not valid in some areas like social sciences.
2. As the order of the partial correlation coefficient goes up, its reliability goes down.
3. The calculations are lengthy and of course, using software solves this problem.

Coefficient of partial correlation This is the correlation coefficient between X_1 and X_2 after the linear effect of X_3 on each of them has been eliminated. The partial correlation coefficient between X_1 and X_2, usually denoted by, $r_{12.3}$ is given by:

$$r_{12.3} = \frac{r_{12} - r_{13}r_{23}}{\sqrt{(1 - r^2{}_{13})(1 - r_{23}{}^2)}}$$

Case Study 11

[7]

Most doctors would probably agree that a Mediterranean diet, rich in vegetables, fruits and grains, is healthier than a high-saturated fat diet. Indeed, previous research

Table 5.17 Description of variables for the Mediterranean diet

Variable	Description
Type of diet	AHA or Mediterranean
Various outcome measures of health and disease	Does the patient have cancer?

Table 5.18 Frequencies for diet and health study

Outcome					
Diet	Cancers	Fatal heart disease	Non-fatal heart disease	Healthy	Total
AHA	15	24	25	239	303
Mediterranean	7	14	8	273	302
Total	22	38	33	512	605

has found that diet can lower the risk of heart disease. However, there is still considerable uncertainty about whether a Mediterranean diet is superior to a low-fat diet recommended by the American Heart Association. This study is the first to compare these two diets.

The subjects, 605 survivors of a heart attack, were randomly assigned to follow either (1) a diet close to the "prudent diet step 1" of the American Heart Association (control group) or (2) a Mediterranean-type diet consisting of more bread and cereals, more fresh fruit and vegetables, more grains, more fish, fewer delicatessen foods, less meat. Experimental canola-oil-based margarine was used instead of butter or cream. The oils recommended for salad and food preparation were canola and olive oils exclusively. Moderate red wine consumption was allowed.

Over four years, patients in the experimental condition were initially seen by the dietician, two months later and then once a year. Compliance with the dietary intervention was checked by a dietary survey and analyses of plasma fatty acids. Patients in the control group were expected to follow the dietary advice given by their physician (Table 5.17).

The researchers collected information on the number of deaths from cardiovascular causes Example: heart attack, strokes, as well as number of nonfatal heart-related episodes. The occurrence of malignant and non-malignant tumours was also carefully monitored.

Contingency Table: It has two variables: diet which is an explanatory variable (in rows) and outcome which is the response variable (in columns).

Frequencies for Diet and Health Study (Table 5.18)

Solution:

Note of Caution: Multivariate analysis is a costly and time-consuming exercise. Multivariate analysis can be performed even for small samples through correlation and regression. Thus, it is better to check on the following points.

1. Multiple correlations, discussed in this chapter, are of linear form. It is very important to check for linearity before applying multiple correlations.
2. It becomes meaningless when there are a large number of independent variables and only a few response variables.
3. The sample under study should be large and representative.

An example to use both the above concepts can be viewed at http://www.sfu.ca/~jackd/Stat203/Wk09_1_Full.pdf.

5.2 Regression

Regression is a mathematical concept that puts together variables that are closely related into an equation that makes impactful decisions with logical sense. This idea was developed by a famous British explorer and anthropologist **Sir Francis Galton**. During his biological experiments, he once observed that the pellets falling formed a normal distribution where the maximum of the bell curve was right below the entry point. See Fig. 5.13. This phenomenon made Galton invent the theory of **simple linear regression**. Regression is a mathematical concept that is used to predict the value of one variable when the other is known. Unlike, correlation the dependent and independent variables are clearly defined here (Fig. 5.14).

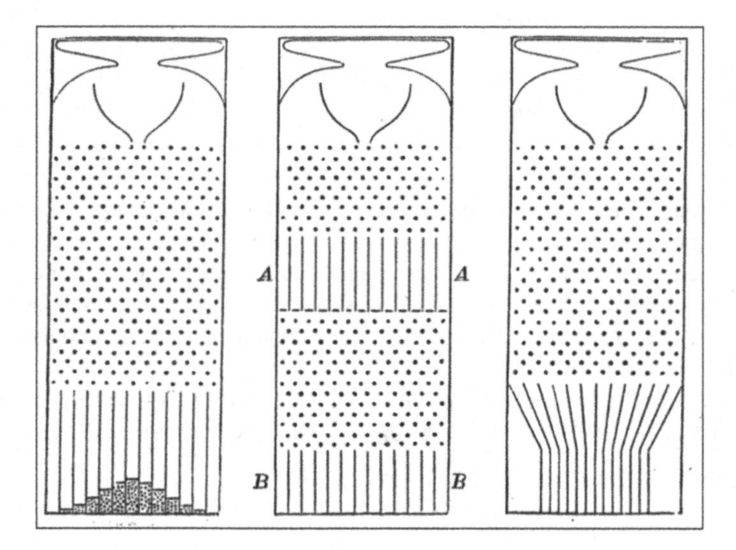

Fig. 5.13 Galton's experiment of pellets falling through a quincunx

Fig. 5.14 Sir Francis Galton

Assumptions for using regression models:

The following assumptions have to be followed before using linear regression models for purposes of accurate prediction:

1. There must be a linear relationship between dependent and independent variables.
2. Independence of the errors. No serial correlations.[1]
3. Normality of the error distribution.
4. Homoscedasticity or constant variance of the errors.
   - Versus time and
   - versus the predictions.

Caution:

[1] When error terms of two subsequent time periods are correlated with each other, it is called serial correlation. For regression models we assume that there is independence of error terms.

1. The regression model is only useful for predicting within the range from which it was estimated. Predicting outside of the range of the model may yield untrustworthy results.
2. Models constructed from too few data points may not be effective or trustworthy.
3. The model should be estimated with a minimum of 10 sample units for every X variable included in a regression equation.

Some commonly asked questions about Regression

Why do we often assume that relationships between variables are *linear*?

Linear relationships are the simplest non-trivial relationships that can be thought of and because the true relationships between variables are often linear over the range of values and most importantly, we can transform the variables to obtain a linear relationship.

Why do we often assume the errors of linear models are normally distributed?

1. It is mathematically convenient. Since we know that there are lots of underlying factors affecting the process and the sum of these individual errors will tend to behave like in a normal distribution.
2. In practice, it seems to be so. Because of the central limit theorem, one can expect even the errors to have mean zero and variance one. This makes modelling data simplified.

Simple Linear Regression

As the name suggests, this is a very basic type of regression with just two quantitative variables defining a relationship between them through a linear equation. The general notation is given as:

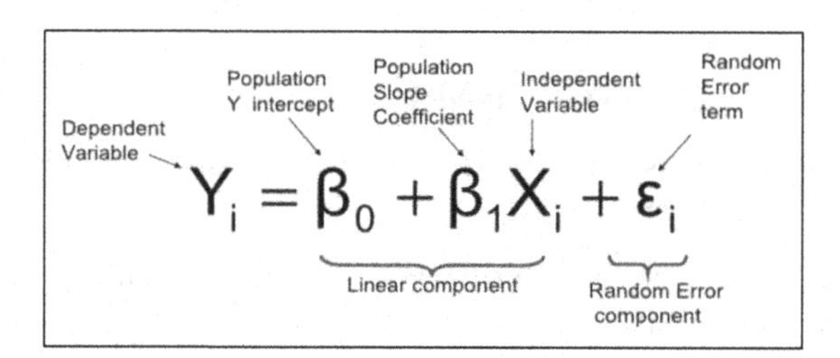

This is a tool to predict the values of one variable from the values of the second variable.

X: Predictor variable or the variable on which the researcher bases his predictions upon.

Y: Criterion variable or the variable we are predicting.

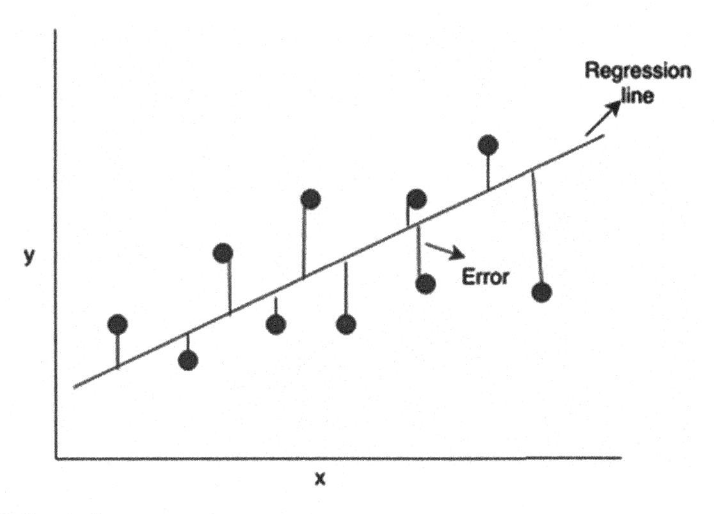

Fig. 5.15 Simple linear regression

When there is only one predictor variable, the prediction method is called simple regression and, in this concept, the predictions of Y when plotted as a function of X form a straight line. See Fig. 5.15.

What do we understand from the linear regression model?

1. What is the level of dependence of the variable X on Y.?
2. Strength of both X and Y variables in the relationship defined.

Multiple Regression

Multiple linear regression MLR is an extension of linear regression models that allow predictions of systems with multiple independent variables. This allows analysts to examine the effect of many different factors on a particular outcome at the same time.

The purpose of multiple regression is to learn more about the relationship between several independent or predictor variables with one dependent variable. This is done by mathematically holding constant all factors but one at a time, the researcher can measure the outcome of a particular factor.

Figure 5.16 represents the 3D view of 3 variables MPG, weight and horsepower of a vehicle which are modelled in a multiple regression model.

Formula:

$$y_i = \beta_o + \beta_1 x_{i1} + \beta_2 x_{i2} + \cdots + \beta_p x_{ip} + \epsilon$$

where, for $i = n$ observations:

y_i Dependent variable.

x_i = Explanatory variable.

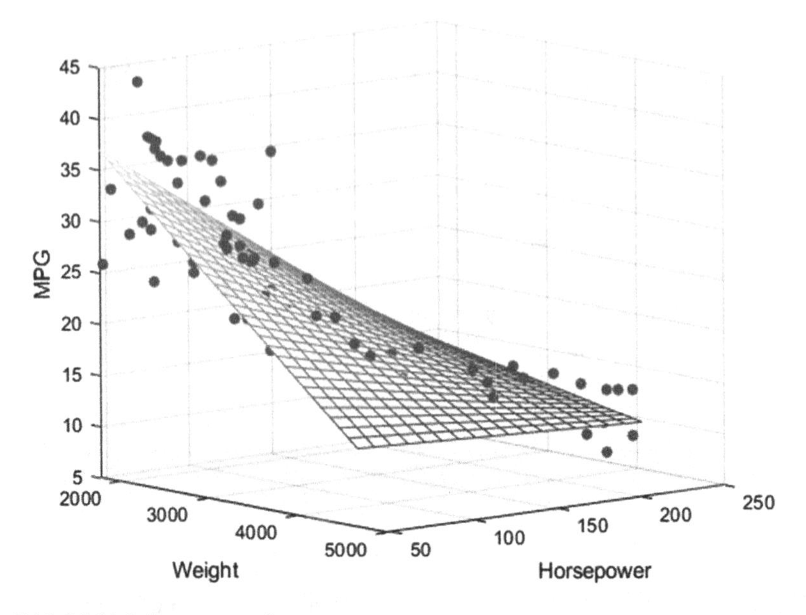

Fig. 5.16 Multiple linear regression

$\beta_0 = Y$ intercept (Constant term)

$\beta_p =$ Slope coefficients for each explanatory variable.

$\epsilon =$ The error term, model's residual.

Logistic Regression

Logistic regression is an extension of simple linear regression like shown in Fig. 5.17.

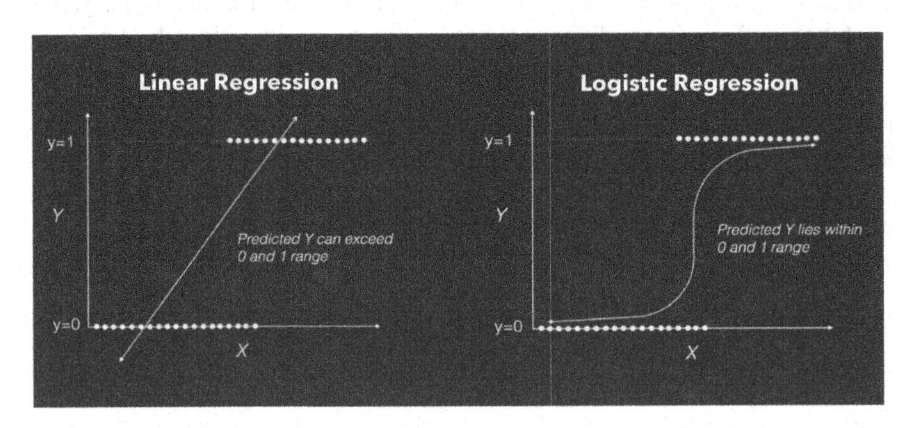

Fig. 5.17 Linear versus logistic regression

When we consider the log of odds of the dependent variable and model it as a linear combination of independent variables, we call it as logistic regression.

When we analyse the data, we must always remember that the output of a logistic regression model is discrete, therefore it is used as a classification tool, whereas linear regression is a model that builds a mathematical equation with related variables so that reliable predictions can be made.

Binary logistic regression This is a special case of logistic regression which is also a classification model, where the dependent variable's response is dichotomous. **Example**: Yes and No, plan as per schedule and plan cancelled. This regression model also predicts the relationship between the independent and dependent variables.

Differences between ANOVA and Regression [8]

1. Both ANOVA and regression seem to be identical, analysts must carefully choose the method based on the type of variables under consideration. Regression is more complicated than ANOVA because there are many concepts in regression and correlation. In ANOVA, people are assigned to treatments but in some areas like social sciences research, we cannot assign people to treatments for practical or ethical reasons.
2. ANOVA generally answers these questions, Are the means of the different data sets equal? If not, then which two data sets seem to be different? Regression on the other hand models the cause and effect of variables under study.
3. So, therefore, the regression equation is a mathematical model used for prediction as it reveals the estimate of an effect, whereas ANOVA measures the mean shift of the responses in different categories of data. See Fig. 5.18, for a better understanding.

Some Research articles using Correlation:

1. **Priory Medical Journals**

Title: Correlations among Depression Rating Scales and a Self-Rating Anxiety Scale in Depressive Outpatients

Authors: Toru Uehara, M.D., Tetsuya Sato, M.D., Kaoru Sakado, M.D.

Illustrates the use of Pearson's correlation.

URL: http://www.priory.com/psych/ratings.html

2. Journal of technology education (Volume 11, Number 1, 1999)

Title: Identification of Quality Characteristics for Technology Education Programs: A North Carolina Case Study

Authors: Aaron C. Clark and Robert E. Wenig

Illustrates the use of Spearman's correlation to assess the strength of the relationship between two sets of ranks.

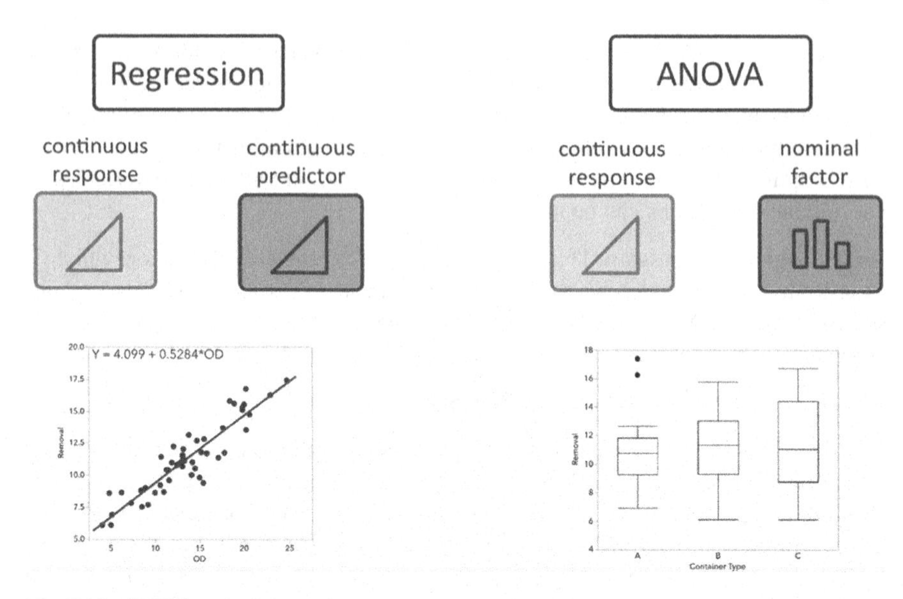

Fig. 5.18 ANOVA versus regression

URL: http://scholar.lib.vt.edu/ejournals/JTE/v11n1/clark.html

Practice data sets on correlation:

1. Data set on: vehicle fuel efficiency from the University of California, Irvine.

URL: https://github.com/AjeetSingh02/ML_correlation_association/blob/master/correlation_association.ipynb

IPython notebook for better understanding the correlation between a mix of categorical and continuous data. This study also utilizes concepts from hypothesis testing, ANOVA and undertakes a few non-parametric tests as well. Correlation analysis was done between

2. Ipython notebook for looking at intercommodity ETF correlations over a set period. Correlation is depicted as a matrix heatmap across various divisions such as agriculture, energy and metals. URL: https://github.com/Commodity-Investing-by-Students/commodity-correlation-notebook/blob/main/COINS_ETF_Correlation.ipynb

References and Research Articles

1. **Correlation comic**: www.dillbert.com/
2. **Spurious correlation**: https://www.tylervigen.com/spurious-correlations

3. **Correlation does not imply causation**: https://towardsdatascience.com/correlation-does-not-imply-causation-92e4832a6713
4. **Correlation between advertisements and sales**: Adapted from: C.A., Inter, 1975.
5. **Study of jobs in Denver neighborhood**: Adopted from neighbourhood facts: The Piton foundation.
6. **Coefficient of correlation**: Adapted from Rajasthan Univ., B. Com 1977.
7. **Mediterranean diet**—De Longerill, M., Salen, P., Martin, J., Monjaud, I., Boucher, P., Mamelle, N. (1998). Mediterranean Dietary pattern in a Randomized Trial. Archives of Internal Medicine, 158, 1181–1187. Is the Mediterranean diet superior to a low-fat diet recommended by the American Heart Association.
8. **Differences between ANOVA and Regression:** https://www.jmp.com/en_in/statistics-knowledge-portal/what-is-regression.html

Statistical Tables

Table A.1 Normal distribution table

z	0.00	0.01	0.02	0.03	0.04	0.05	0.06	0.07	0.08	0.09
0.0	0.5000	0.5040	0.5080	0.5120	0.5160	0.5199	0.5239	0.5279	0.5319	0.5359
0.1	0.5398	0.5438	0.5478	0.5517	0.5557	0.5596	0.5636	0.5675	0.5714	0.5753
0.2	0.5793	0.5832	0.5871	0.5910	0.5948	0.5987	0.6026	0.6064	0.6103	0.6141
0.3	0.6179	0.6217	0.6255	0.6293	0.6331	0.6368	0.6406	0.6443	0.6480	0.6517
0.4	0.6554	0.6591	0.6628	0.6664	0.6700	0.6736	0.6772	0.6808	0.6844	0.6879
0.5	0.6915	0.6950	0.6985	0.7019	0.7054	0.7088	0.7123	0.7157	0.7190	0.7224
0.6	0.7257	0.7291	0.7324	0.7357	0.7389	0.7422	0.7454	0.7486	0.7517	0.7549
0.7	0.7580	0.7611	0.7642	0.7673	0.7704	0.7734	0.7764	0.7794	0.7823	0.7852
0.8	0.7881	0.7910	0.7939	0.7967	0.7995	0.8023	0.8051	0.8078	0.8106	0.8133
0.9	0.8159	0.8186	0.8212	0.8238	0.8264	0.8289	0.8315	0.8340	0.8365	0.8389
1.0	0.8413	0.8438	0.8461	0.8485	0.8508	0.8531	0.8554	0.8577	0.8599	0.8621
1.1	0.8643	0.8665	0.8686	0.8708	0.8729	0.8749	0.8770	0.8790	0.8810	0.8830
1.2	0.8849	0.8869	0.8888	0.8907	0.8925	0.8944	0.8962	0.8980	0.8997	0.9015
1.3	0.9032	0.9049	0.9066	0.9082	0.9099	0.9115	0.9131	0.9147	0.9162	0.9177
1.4	0.9192	0.9207	0.9222	0.9236	0.9251	0.9265	0.9279	0.9292	0.9306	0.9319
1.5	0.9332	0.9345	0.9357	0.9370	0.9382	0.9394	0.9406	0.9418	0.9429	0.9441
1.6	0.9452	0.9463	0.9474	0.9484	0.9495	0.9505	0.9515	0.9525	0.9535	0.9545
1.7	0.9554	0.9564	0.9573	0.9582	0.9591	0.9599	0.9608	0.9616	0.9625	0.9633
1.8	0.9641	0.9649	0.9656	0.9664	0.9671	0.9678	0.9686	0.9693	0.9699	0.9706
1.9	0.9713	0.9719	0.9726	0.9732	0.9738	0.9744	0.9750	0.9756	0.9761	0.9767
2.0	0.9772	0.9778	0.9783	0.9788	0.9793	0.9798	0.9803	0.9808	0.9812	0.9817
2.1	0.9821	0.9826	0.9830	0.9834	0.9838	0.9842	0.9846	0.9850	0.9854	0.9857
2.2	0.9861	0.9864	0.9868	0.9871	0.9875	0.9878	0.9881	0.9884	0.9887	0.9890

(continued)

S. Prasad, *Elementary Statistical Methods*,
https://doi.org/10.1007/978-981-19-0596-4

(continued)

z	0.00	0.01	0.02	0.03	0.04	0.05	0.06	0.07	0.08	0.09
2.3	0.9893	0.9896	0.9898	0.9901	0.9904	0.9906	0.9909	0.9911	0.9913	0.9916
2.4	0.9918	0.9920	0.9922	0.9925	0.9927	0.9929	0.9931	0.9932	0.9934	0.9936
2.5	0.9938	0.9940	0.9941	0.9943	0.9945	0.9946	0.9948	0.9949	0.9951	0.9952
2.6	0.9953	0.9955	0.9956	0.9957	0.9959	0.9960	0.9961	0.9962	0.9963	0.9964
2.7	0.9965	0.9966	0.9967	0.9968	0.9969	0.9970	0.9971	0.9972	0.9973	0.9974
2.8	0.9974	0.9975	0.9976	0.9977	0.9977	0.9978	0.9979	0.9979	0.9980	0.9981
2.9	0.9981	0.9982	0.9982	0.9983	0.9984	0.9984	0.9985	0.9985	0.9986	0.9986
3.0	0.9987	0.9987	0.9987	0.9988	0.9988	0.9989	0.9989	0.9989	0.9990	0.9990
3.1	0.9990	0.9991	0.9991	0.9991	0.9992	0.9992	0.9992	0.9992	0.9993	0.9993
3.2	0.9993	0.9993	0.9994	0.9994	0.9994	0.9994	0.9994	0.9995	0.9995	0.9995
3.3	0.9995	0.9995	0.9995	0.9996	0.9996	0.9996	0.9996	0.9996	0.9996	0.9997
3.4	0.9997	0.9997	0.9997	0.9997	0.9997	0.9997	0.9997	0.9997	0.9997	0.9998
3.5	0.9998	0.9998	0.9998	0.9998	0.9998	0.9998	0.9998	0.9998	0.9998	0.9998
3.6	0.9998	0.9998	0.9999							

Table A.2 *t*-distribution table

	Significance level					
Degrees of freedom	10%	5%	2%	1%	0.2%	0.1%
	5%	2.5%	1%	0.5%	0.1%	0.05%
1	6.314	12.706	31.821	63.657	318.309	636.619
2	2.920	4.303	6.965	9.925	22.327	31.599
3	2.353	3.182	4.541	5.841	10.215	12.924
4	2.132	2.776	3.747	4.604	7.173	8.610
5	2.015	2.571	3.365	4.032	5.893	6.869
6	1.943	2.447	3.143	3.707	5.208	5.959
7	1.894	2.365	2.998	3.499	4.785	5.408
8	1.860	2.306	2.896	3.355	4.501	5.041
9	1.833	2.262	2.821	3.250	4.297	4.781
10	1.812	2.228	2.764	3.169	4.144	4.587
11	1.796	2.201	2.718	3.106	4.025	4.437
12	1.782	2.179	2.681	3.055	3.930	4.318
13	1.771	2.160	2.650	3.012	3.852	4.221
14	1.761	2.145	2.624	2.977	3.787	4.140
15	1.753	2.131	2.602	2.947	3.733	4.073
16	1.746	2.120	2.583	2.921	3.686	4.015
17	1.740	2.110	2.567	2.898	3.646	3.965
18	1.734	2.101	2.552	2.878	3.610	3.922
19	1.729	2.093	2.539	2.861	3.579	3.883
20	1.725	2.086	2.528	2.845	3.552	3.850
21	1.721	2.080	2.518	2.831	3.527	3.819
22	1.717	2.074	2.508	2.819	3.505	3.792
23	1.714	2.069	2.500	2.807	3.485	3.768
24	1.711	2.064	2.492	2.797	3.467	3.745
25	1.708	2.060	2.485	2.787	3.450	3.725
26	1.706	2.056	2.479	2.779	3.435	3.707
27	1.703	2.052	2.473	2.771	3.421	3.690
28	1.701	2.048	2.467	2.763	3.408	3.674
29	1.699	2.045	2.462	2.756	3.396	3.659
30	1.697	2.042	2.457	2.750	3.385	3.646
32	1.694	2.037	2.449	2.738	3.365	3.622
34	1.691	2.032	2.441	2.728	3.348	3.601
36	1.688	2.028	2.434	2.719	3.333	3.582
38	1.686	2.024	2.429	2.712	3.319	3.566

(continued)

Table A.2 (continued)

	Significance level					
Degrees of freedom	10%	5%	2%	1%	0.2%	0.1%
	5%	2.5%	1%	0.5%	0.1%	0.05%
40	1.684	2.021	2.423	2.704	3.307	3.551
42	1.682	2.018	2.418	2.698	3.296	3.538
44	1.680	2.015	2.414	2.692	3.286	3.526
46	1.679	2.013	2.410	2.687	3.277	3.515
48	1.677	2.011	2.407	2.682	3.269	3.505
50	1.676	2.009	2.403	2.678	3.261	3.496
60	1.671	2.000	2.390	2.660	3.232	3.460
70	1.667	1.994	2.381	2.648	3.211	3.435
80	1.664	1.990	2.374	2.639	3.195	3.416
90	1.662	1.987	2.368	2.632	3.183	3.402
100	1.660	1.984	2.364	2.626	3.174	3.390
120	1.658	1.980	2.358	2.617	3.160	3.373
150	1.655	1.976	2.351	2.609	3.145	3.357
200	1.653	1.972	2.345	2.601	3.131	3.340
300	1.650	1.968	2.339	2.592	3.118	3.323
400	1.649	1.966	2.336	2.588	3.111	3.315
500	1.648	1.965	2.334	2.586	3.107	3.310
600	1.647	1.964	2.333	2.584	3.104	3.307
∞	1.645	1.960	2.326	2.576	3.090	3.291

Table A.3 *F*-distribution table

	ν_1														
ν_2	1	2	3	4	5	6	7	8	9	10	12	14	16	18	20
1	161.45	199.50	215.71	224.58	230.16	233.99	236.77	238.88	240.54	241.88	243.91	245.36	246.46	247.32	248.01
2	18.51	19.00	19.16	19.25	19.30	19.33	19.35	19.37	19.38	19.40	19.41	19.42	19.43	19.44	19.45
3	10.13	9.55	9.28	9.12	9.01	8.94	8.89	8.85	8.81	8.79	8.74	8.71	8.69	8.67	8.66
4	7.71	6.94	6.59	6.39	6.26	6.16	6.09	6.04	6.00	5.96	5.91	5.87	5.84	5.82	5.80
5	6.61	5.79	5.41	5.19	5.05	4.95	4.88	4.82	4.77	4.74	4.68	4.64	4.60	4.58	4.56
6	5.99	5.14	4.76	4.53	4.39	4.28	4.21	4.15	4.10	4.06	4.00	3.96	3.92	3.90	3.87
7	5.59	4.74	4.35	4.12	3.97	3.87	3.79	3.73	3.68	3.64	3.57	3.53	3.49	3.47	3.44
8	5.32	4.46	4.07	3.84	3.69	3.58	3.50	3.44	3.39	3.35	3.28	3.24	3.20	3.17	3.15
9	5.12	4.26	3.86	3.63	3.48	3.37	3.29	3.23	3.18	3.14	3.07	3.03	2.99	2.96	2.94
10	4.96	4.10	3.71	3.48	3.33	3.22	3.14	3.07	3.02	2.98	2.91	2.86	2.83	2.80	2.77
11	4.84	3.98	3.59	3.36	3.20	3.09	3.01	2.95	2.90	2.85	2.79	2.74	2.70	2.67	2.65
12	4.75	3.89	3.49	3.26	3.11	3.00	2.91	2.85	2.80	2.75	2.69	2.64	2.60	2.57	2.54
13	4.67	3.81	3.41	3.18	3.03	2.92	2.83	2.77	2.71	2.67	2.60	2.55	2.51	2.48	2.46
14	4.60	3.74	3.34	3.11	2.96	2.85	2.76	2.70	2.65	2.60	2.53	2.48	2.44	2.41	2.39
15	4.54	3.68	3.29	3.06	2.90	2.79	2.71	2.64	2.59	2.54	2.48	2.42	2.38	2.35	2.33
16	4.49	3.63	3.24	3.01	2.85	2.74	2.66	2.59	2.54	2.49	2.42	2.37	2.33	2.30	2.28
17	4.45	3.59	3.20	2.96	2.81	2.70	2.61	2.55	2.49	2.45	2.38	2.33	2.29	2.26	2.23
18	4.41	3.55	3.16	2.93	2.77	2.66	2.58	2.51	2.46	2.41	2.34	2.29	2.25	2.22	2.19
19	4.38	3.52	3.13	2.90	2.74	2.63	2.54	2.48	2.42	2.38	2.31	2.26	2.21	2.18	2.16
20	4.35	3.49	3.10	2.87	2.71	2.60	2.51	2.45	2.39	2.35	2.28	2.22	2.18	2.15	2.12

(continued)

Table A.3 (continued)

	v_1														
v_2	1	2	3	4	5	6	7	8	9	10	12	14	16	18	20
21	4.32	3.47	3.07	2.84	2.68	2.57	2.49	2.42	2.37	2.32	2.25	2.20	2.16	2.12	2.10
22	4.30	3.44	3.05	2.82	2.66	2.55	2.46	2.40	2.34	2.30	2.23	2.17	2.13	2.10	2.07
23	4.28	3.42	3.03	2.80	2.64	2.53	2.44	2.37	2.32	2.27	2.20	2.15	2.11	2.08	2.05
24	4.26	3.40	3.01	2.78	2.62	2.51	2.42	2.36	2.30	2.25	2.18	2.13	2.09	2.05	2.03
25	4.24	3.39	2.99	2.76	2.60	2.49	2.40	2.34	2.28	2.24	2.16	2.11	2.07	2.04	2.01
26	4.22	3.37	2.98	2.74	2.59	2.47	2.39	2.32	2.27	2.22	2.15	2.09	2.05	2.02	1.99
27	4.21	3.35	2.96	2.73	2.57	2.46	2.37	2.31	2.25	2.20	2.13	2.08	2.04	2.00	1.97
28	4.20	3.34	2.95	2.71	2.56	2.45	2.36	2.29	2.24	2.19	2.12	2.06	2.02	1.99	1.96
29	4.18	3.33	2.93	2.70	2.55	2.43	2.35	2.28	2.22	2.18	2.10	2.05	2.01	1.97	1.94
30	4.17	3.32	2.92	2.69	2.53	2.42	2.33	2.27	2.21	2.16	2.09	2.04	1.99	1.96	1.93
35	4.12	3.27	2.87	2.64	2.49	2.37	2.29	2.22	2.16	2.11	2.04	1.99	1.94	1.91	1.88
40	4.08	3.23	2.84	2.61	2.45	2.34	2.25	2.18	2.12	2.08	2.00	1.95	1.90	1.87	1.84
50	4.03	3.18	2.79	2.56	2.40	2.29	2.20	2.13	2.07	2.03	1.95	1.89	1.85	1.81	1.78
60	4.00	3.15	2.76	2.53	2.37	2.25	2.17	2.10	2.04	1.99	1.92	1.86	1.82	1.78	1.75
70	3.98	3.13	2.74	2.50	2.35	2.23	2.14	2.07	2.02	1.97	1.89	1.84	1.79	1.75	1.72
80	3.96	3.11	2.72	2.49	2.33	2.21	2.13	2.06	2.00	1.95	1.88	1.82	1.77	1.73	1.70
90	3.95	3.10	2.71	2.47	2.32	2.20	2.11	2.04	1.99	1.94	1.86	1.80	1.76	1.72	1.69
100	3.94	3.09	2.70	2.46	2.31	2.19	2.10	2.03	1.97	1.93	1.85	1.79	1.75	1.71	1.68
120	3.92	3.07	2.68	2.45	2.29	2.18	2.09	2.02	1.96	1.91	1.83	1.78	1.73	1.69	1.66

(continued)

Table A.3 (continued)

	v_1														
v_2	1	2	3	4	5	6	7	8	9	10	12	14	16	18	20
150	3.90	3.06	2.66	2.43	2.27	2.16	2.07	2.00	1.94	1.89	1.82	1.76	1.71	1.67	1.64
200	3.89	3.04	2.65	2.42	2.26	2.14	2.06	1.98	1.93	1.88	1.80	1.74	1.69	1.66	1.62
250	3.88	3.03	2.64	2.41	2.25	2.13	2.05	1.98	1.92	1.87	1.79	1.73	1.68	1.65	1.61
300	3.87	3.03	2.63	2.40	2.24	2.13	2.04	1.97	1.91	1.86	1.78	1.72	1.68	1.64	1.61
400	3.86	3.02	2.63	2.39	2.24	2.12	2.03	1.96	1.90	1.85	1.78	1.72	1.67	1.63	1.60
500	3.86	3.01	2.62	2.39	2.23	2.12	2.03	1.96	1.90	1.85	1.77	1.71	1.66	1.62	1.59
600	3.86	3.01	2.62	2.39	2.23	2.11	2.02	1.95	1.90	1.85	1.77	1.71	1.66	1.62	1.59
750	3.85	3.01	2.62	2.38	2.23	2.11	2.02	1.95	1.89	1.84	1.77	1.70	1.66	1.62	1.58
1000	3.85	3.00	2.61	2.38	2.22	2.11	2.02	1.95	1.89	1.84	1.76	1.70	1.65	1.61	1.58

	v_1									
v_2	25	30	35	40	50	65	75	100	150	200
1	249.26	250.10	250.69	251.14	251.77	252.20	252.62	253.04	253.46	253.68
2	19.46	19.46	19.47	19.47	19.48	19.48	19.48	19.49	19.49	19.49
3	8.63	8.62	8.60	8.59	8.58	8.57	8.56	8.55	8.54	8.54
4	5.77	5.75	5.73	5.72	5.70	5.69	5.68	5.66	5.65	5.65
5	4.52	4.50	4.48	4.46	4.44	4.43	4.42	4.41	4.39	4.39
6	3.83	3.81	3.79	3.77	3.75	3.74	3.73	3.71	3.70	3.69
7	3.40	3.38	3.36	3.34	3.32	3.30	3.29	3.27	3.26	3.25
8	3.11	3.08	3.06	3.04	3.02	3.01	2.99	2.97	2.96	2.95

(continued)

Table A.3 (continued)

	v_1									
v_2	25	30	35	40	50	65	75	100	150	200
9	2.89	2.86	2.84	2.83	2.80	2.79	2.77	2.76	2.74	2.73
10	2.73	2.70	2.68	2.66	2.64	2.62	2.60	2.59	2.57	2.56
11	2.60	2.57	2.55	2.53	2.51	2.49	2.47	2.46	2.44	2.43
12	2.50	2.47	2.44	2.43	2.40	2.38	2.37	2.35	2.33	2.32
13	2.41	2.38	2.36	2.34	2.31	2.30	2.28	2.26	2.24	2.23
14	2.34	2.31	2.28	2.27	2.24	2.22	2.21	2.19	2.17	2.16
15	2.28	2.25	2.22	2.20	2.18	2.16	2.14	2.12	2.10	2.10
16	2.23	2.19	2.17	2.15	2.12	2.11	2.09	2.07	2.05	2.04
17	2.18	2.15	2.12	2.10	2.08	2.06	2.04	2.02	2.00	1.99
18	2.14	2.11	2.08	2.06	2.04	2.02	2.00	1.98	1.96	1.95
19	2.11	2.07	2.05	2.03	2.00	1.98	1.96	1.94	1.92	1.91
20	2.07	2.04	2.01	1.99	1.97	1.95	1.93	1.91	1.89	1.88
21	2.05	2.01	1.98	1.96	1.94	1.92	1.90	1.88	1.86	1.84
22	2.02	1.98	1.96	1.94	1.91	1.89	1.87	1.85	1.83	1.82
23	2.00	1.96	1.93	1.91	1.88	1.86	1.84	1.82	1.80	1.79
24	1.97	1.94	1.91	1.89	1.86	1.84	1.82	1.80	1.78	1.77
25	1.96	1.92	1.89	1.87	1.84	1.82	1.80	1.78	1.76	1.75
26	1.94	1.90	1.87	1.85	1.82	1.80	1.78	1.76	1.74	1.73
27	1.92	1.88	1.86	1.84	1.81	1.79	1.76	1.74	1.72	1.71

(continued)

Table A.3 (continued)

v_2	v_1 25	30	35	40	50	65	75	100	150	200
28	1.91	1.87	1.84	1.82	1.79	1.77	1.75	1.73	1.70	1.69
29	1.89	1.85	1.83	1.81	1.77	1.75	1.73	1.71	1.69	1.67
30	1.88	1.84	1.81	1.79	1.76	1.74	1.72	1.70	1.67	1.66
35	1.82	1.79	1.76	1.74	1.70	1.68	1.66	1.63	1.61	1.60
40	1.78	1.74	1.72	1.69	1.66	1.64	1.61	1.59	1.56	1.55
50	1.73	1.69	1.66	1.63	1.60	1.58	1.55	1.52	1.50	1.48
60	1.69	1.65	1.62	1.59	1.56	1.53	1.51	1.48	1.45	1.44
70	1.66	1.62	1.59	1.57	1.53	1.50	1.48	1.45	1.42	1.40
80	1.64	1.60	1.57	1.54	1.51	1.48	1.45	1.43	1.39	1.38
90	1.63	1.59	1.55	1.53	1.49	1.46	1.44	1.41	1.38	1.36
100	1.62	1.57	1.54	1.52	1.48	1.45	1.42	1.39	1.36	1.34
120	1.60	1.55	1.52	1.50	1.46	1.43	1.40	1.37	1.33	1.32
150	1.58	1.54	1.50	1.48	1.44	1.41	1.38	1.34	1.31	1.29
200	1.56	1.52	1.48	1.46	1.41	1.39	1.35	1.32	1.28	1.26
250	1.55	1.50	1.47	1.44	1.40	1.37	1.34	1.31	1.27	1.25
300	1.54	1.50	1.46	1.43	1.39	1.36	1.33	1.30	1.26	1.23
400	1.53	1.49	1.45	1.42	1.38	1.35	1.32	1.28	1.24	1.22
500	1.53	1.48	1.45	1.42	1.38	1.35	1.31	1.28	1.23	1.21
600	1.52	1.48	1.44	1.41	1.37	1.34	1.31	1.27	1.23	1.20

(continued)

Table A.3 (continued)

v_2	v_1 25	30	35	40	50	65	75	100	150	200
750	1.52	1.47	1.44	1.41	1.37	1.34	1.30	1.26	1.22	1.20
1000	1.52	1.47	1.43	1.41	1.36	1.33	1.30	1.26	1.22	1.19

v_2	v_1 1	2	3	4	5	6	7	8	9	10	11	12	13	14	15
1	4052.18	4999.50	5403.35	5624.58	5763.65	5858.99	5928.36	5981.07	6022.47	6055.85	6106.32	6142.67	6170.10	6191.53	6208.73
2	98.50	99.00	99.17	99.25	99.30	99.33	99.36	99.37	99.39	99.40	99.42	99.43	99.44	99.44	99.45
3	34.12	30.82	29.46	28.71	28.24	27.91	27.67	27.49	27.35	27.23	27.05	26.92	26.83	26.75	26.69
4	21.20	18.00	16.69	15.98	15.52	15.21	14.98	14.80	14.66	14.55	14.37	14.25	14.15	14.08	14.02
5	16.26	13.27	12.06	11.39	10.97	10.67	10.46	10.29	10.16	10.05	9.89	9.77	9.68	9.61	9.55
6	13.75	10.92	9.78	9.15	8.75	8.47	8.26	8.10	7.98	7.87	7.72	7.60	7.52	7.45	7.40
7	12.25	9.55	8.45	7.85	7.46	7.19	6.99	6.84	6.72	6.62	6.47	6.36	6.28	6.21	6.16
8	11.26	8.65	7.59	7.01	6.63	6.37	6.18	6.03	5.91	5.81	5.67	5.56	5.48	5.41	5.36
9	10.56	8.02	6.99	6.42	6.06	5.80	5.61	5.47	5.35	5.26	5.11	5.01	4.92	4.86	4.81
10	10.04	7.56	6.55	5.99	5.64	5.39	5.20	5.06	4.94	4.85	4.71	4.60	4.52	4.46	4.41
11	9.65	7.21	6.22	5.67	5.32	5.07	4.89	4.74	4.63	4.54	4.40	4.29	4.21	4.15	4.10
12	9.33	6.93	5.95	5.41	5.06	4.82	4.64	4.50	4.39	4.30	4.16	4.05	3.97	3.91	3.86
13	9.07	6.70	5.74	5.21	4.86	4.62	4.44	4.30	4.19	4.10	3.96	3.86	3.78	3.72	3.66
14	8.86	6.51	5.56	5.04	4.69	4.46	4.28	4.14	4.03	3.94	3.80	3.70	3.62	3.56	3.51
15	8.68	6.36	5.42	4.89	4.56	4.32	4.14	4.00	3.89	3.80	3.67	3.56	3.49	3.42	3.37

(continued)

Table A.3 (continued)

ν_2 \ ν_1	1	2	3	4	5	6	7	8	9	10	11	12	13	14	15
16	8.53	6.23	5.29	4.77	4.44	4.20	4.03	3.89	3.78	3.69	3.55	3.45	3.37	3.31	3.26
17	8.40	6.11	5.18	4.67	4.34	4.10	3.93	3.79	3.68	3.59	3.46	3.35	3.27	3.21	3.16
18	8.29	6.01	5.09	4.58	4.25	4.01	3.84	3.71	3.60	3.51	3.37	3.27	3.19	3.13	3.08
19	8.18	5.93	5.01	4.50	4.17	3.94	3.77	3.63	3.52	3.43	3.30	3.19	3.12	3.05	3.00
20	8.10	5.85	4.94	4.43	4.10	3.87	3.70	3.56	3.46	3.37	3.23	3.13	3.05	2.99	2.94
21	8.02	5.78	4.87	4.37	4.04	3.81	3.64	3.51	3.40	3.31	3.17	3.07	2.99	2.93	2.88
22	7.95	5.72	4.82	4.31	3.99	3.76	3.59	3.45	3.35	3.26	3.12	3.02	2.94	2.88	2.83
23	7.88	5.66	4.76	4.26	3.94	3.71	3.54	3.41	3.30	3.21	3.07	2.97	2.89	2.83	2.78
24	7.82	5.61	4.72	4.22	3.90	3.67	3.50	3.36	3.26	3.17	3.03	2.93	2.85	2.79	2.74
25	7.77	5.57	4.68	4.18	3.85	3.63	3.46	3.32	3.22	3.13	2.99	2.89	2.81	2.75	2.70
26	7.72	5.53	4.64	4.14	3.82	3.59	3.42	3.29	3.18	3.09	2.96	2.86	2.78	2.72	2.66
27	7.68	5.49	4.60	4.11	3.78	3.56	3.39	3.26	3.15	3.06	2.93	2.82	2.75	2.68	2.63
28	7.64	5.45	4.57	4.07	3.75	3.53	3.36	3.23	3.12	3.03	2.90	2.79	2.72	2.65	2.60
29	7.60	5.42	4.54	4.04	3.73	3.50	3.33	3.20	3.09	3.00	2.87	2.77	2.69	2.63	2.57
30	7.56	5.39	4.51	4.02	3.70	3.47	3.30	3.17	3.07	2.98	2.84	2.74	2.66	2.60	2.55
35	7.42	5.27	4.40	3.91	3.59	3.37	3.20	3.07	2.96	2.88	2.74	2.64	2.56	2.50	2.44
40	7.31	5.18	4.31	3.83	3.51	3.29	3.12	2.99	2.89	2.80	2.66	2.56	2.48	2.42	2.37
50	7.17	5.06	4.20	3.72	3.41	3.19	3.02	2.89	2.78	2.70	2.56	2.46	2.38	2.32	2.27
60	7.08	4.98	4.13	3.65	3.34	3.12	2.95	2.82	2.72	2.63	2.50	2.39	2.31	2.25	2.20

(continued)

Table A.3 (continued)

	ν_1														
ν_2	1	2	3	4	5	6	7	8	9	10	11	12	13	14	15
70	7.01	4.92	4.07	3.60	3.29	3.07	2.91	2.78	2.67	2.59	2.45	2.35	2.27	2.20	2.15
80	6.96	4.88	4.04	3.56	3.26	3.04	2.87	2.74	2.64	2.55	2.42	2.31	2.23	2.17	2.12
90	6.93	4.85	4.01	3.53	3.23	3.01	2.84	2.72	2.61	2.52	2.39	2.29	2.21	2.14	2.09
100	6.90	4.82	3.98	3.51	3.21	2.99	2.82	2.69	2.59	2.50	2.37	2.27	2.19	2.12	2.07
120	6.85	4.79	3.95	3.48	3.17	2.96	2.79	2.66	2.56	2.47	2.34	2.23	2.15	2.09	2.03
150	6.81	4.75	3.91	3.45	3.14	2.92	2.76	2.63	2.53	2.44	2.31	2.20	2.12	2.06	2.00
200	6.76	4.71	3.88	3.41	3.11	2.89	2.73	2.60	2.50	2.41	2.27	2.17	2.09	2.03	1.97
250	6.74	4.69	3.86	3.40	3.09	2.87	2.71	2.58	2.48	2.39	2.26	2.15	2.07	2.01	1.95
300	6.72	4.68	3.85	3.38	3.08	2.86	2.70	2.57	2.47	2.38	2.24	2.14	2.06	1.99	1.94
400	6.70	4.66	3.83	3.37	3.06	2.85	2.68	2.56	2.45	2.37	2.23	2.13	2.05	1.98	1.92
500	6.69	4.65	3.82	3.36	3.05	2.84	2.68	2.55	2.44	2.36	2.22	2.12	2.04	1.97	1.92
600	6.68	4.64	3.81	3.35	3.05	2.83	2.67	2.54	2.44	2.35	2.21	2.11	2.03	1.96	1.91
750	6.67	4.63	3.81	3.34	3.04	2.83	2.66	2.53	2.43	2.34	2.21	2.11	2.02	1.96	1.90
1000	6.66	4.63	3.80	3.34	3.04	2.82	2.66	2.53	2.43	2.34	2.20	2.10	2.02	1.95	1.90

	ν_1									
ν_2	25	30	35	40	50	60	75	100	150	200
1	6239.83	6260.65	6275.57	6286.78	6302.52	6313.03	6323.56	6334.11	6344.68	6349.97
2	99.46	99.47	99.47	99.47	99.48	99.48	99.49	99.49	99.49	99.49
3	26.58	26.50	26.45	26.41	26.35	26.32	26.28	26.24	26.20	26.18

(continued)

Table A.3 (continued)

	ν_1									
ν_2	25	30	35	40	50	60	75	100	150	200
4	13.91	13.84	13.79	13.75	13.69	13.65	13.61	13.58	13.54	13.52
5	9.45	9.38	9.33	9.29	9.24	9.20	9.17	9.13	9.09	9.08
6	7.30	7.23	7.18	7.14	7.09	7.06	7.02	6.99	6.95	6.93
7	6.06	5.99	5.94	5.91	5.86	5.82	5.79	5.75	5.72	5.70
8	5.26	5.20	5.15	5.12	5.07	5.03	5.00	4.96	4.93	4.91
9	4.71	4.65	4.60	4.57	4.52	4.48	4.45	4.41	4.38	4.36
10	4.31	4.25	4.20	4.17	4.12	4.08	4.05	4.01	3.98	3.96
11	4.01	3.94	3.89	3.86	3.81	3.78	3.74	3.71	3.67	3.66
12	3.76	3.70	3.65	3.62	3.57	3.54	3.50	3.47	3.43	3.41
13	3.57	3.51	3.46	3.43	3.38	3.34	3.31	3.27	3.24	3.22
14	3.41	3.35	3.30	3.27	3.22	3.18	3.15	3.11	3.08	3.06
15	3.28	3.21	3.17	3.13	3.08	3.05	3.01	2.98	2.94	2.92
16	3.16	3.10	3.05	3.02	2.97	2.93	2.90	2.86	2.83	2.81
17	3.07	3.00	2.96	2.92	2.87	2.83	2.80	2.76	2.73	2.71
18	2.98	2.92	2.87	2.84	2.78	2.75	2.71	2.68	2.64	2.62
19	2.91	2.84	2.80	2.76	2.71	2.67	2.64	2.60	2.57	2.55
20	2.84	2.78	2.73	2.69	2.64	2.61	2.57	2.54	2.50	2.48
21	2.79	2.72	2.67	2.64	2.58	2.55	2.51	2.48	2.44	2.42
22	2.73	2.67	2.62	2.58	2.53	2.50	2.46	2.42	2.38	2.36

(continued)

Table A.3 (continued)

	ν_1									
ν_2	25	30	35	40	50	60	75	100	150	200
23	2.69	2.62	2.57	2.54	2.48	2.45	2.41	2.37	2.34	2.32
24	2.64	2.58	2.53	2.49	2.44	2.40	2.37	2.33	2.29	2.27
25	2.60	2.54	2.49	2.45	2.40	2.36	2.33	2.29	2.25	2.23
26	2.57	2.50	2.45	2.42	2.36	2.33	2.29	2.25	2.21	2.19
27	2.54	2.47	2.42	2.38	2.33	2.29	2.26	2.22	2.18	2.16
28	2.51	2.44	2.39	2.35	2.30	2.26	2.23	2.19	2.15	2.13
29	2.48	2.41	2.36	2.33	2.27	2.23	2.20	2.16	2.12	2.10
30	2.45	2.39	2.34	2.30	2.25	2.21	2.17	2.13	2.09	2.07
35	2.35	2.28	2.23	2.19	2.14	2.10	2.06	2.02	1.98	1.96
40	2.27	2.20	2.15	2.11	2.06	2.02	1.98	1.94	1.90	1.87
50	2.17	2.10	2.05	2.01	1.95	1.91	1.87	1.82	1.78	1.76
60	2.10	2.03	1.98	1.94	1.88	1.84	1.79	1.75	1.70	1.68
70	2.05	1.98	1.93	1.89	1.83	1.78	1.74	1.70	1.65	1.62
80	2.01	1.94	1.89	1.85	1.79	1.75	1.70	1.65	1.61	1.58
90	1.99	1.92	1.86	1.82	1.76	1.72	1.67	1.62	1.57	1.55
100	1.97	1.89	1.84	1.80	1.74	1.69	1.65	1.60	1.55	1.52
120	1.93	1.86	1.81	1.76	1.70	1.66	1.61	1.56	1.51	1.48
150	1.90	1.83	1.77	1.73	1.66	1.62	1.57	1.52	1.46	1.43
200	1.87	1.79	1.74	1.69	1.63	1.58	1.53	1.48	1.42	1.39

(continued)

Table A.3 (continued)

	v_1									
v_2	25	30	35	40	50	60	75	100	150	200
250	1.85	1.77	1.72	1.67	1.61	1.56	1.51	1.46	1.40	1.36
300	1.84	1.76	1.70	1.66	1.59	1.55	1.50	1.44	1.38	1.35
400	1.82	1.75	1.69	1.64	1.58	1.53	1.48	1.42	1.36	1.32
500	1.81	1.74	1.68	1.63	1.57	1.52	1.47	1.41	1.34	1.31
600	1.80	1.73	1.67	1.63	1.56	1.51	1.46	1.40	1.34	1.30
750	1.80	1.72	1.66	1.62	1.55	1.50	1.45	1.39	1.33	1.29
1000	1.79	1.72	1.66	1.61	1.54	1.50	1.44	1.38	1.32	1.28

	v_1														
v_2	1	2	3	4	5	6	7	8	9	10	12	14	16	18	20
1	4.05e05	5.00e05	5.40e05	5.62e05	5.76e05	5.86e05	5.93e05	5.98e05	6.02e05	6.06e05	6.11e05	6.14e05	6.17e05	6.19e05	6.21e05
2	998.50	999.00	999.17	999.25	999.30	999.33	999.36	999.37	999.39	999.40	999.42	999.43	999.44	999.44	999.45
3	167.03	148.50	141.11	137.10	134.58	132.85	131.58	130.62	129.86	129.25	128.32	127.64	127.14	126.74	126.42
4	74.14	61.25	56.18	53.44	51.71	50.53	49.66	49.00	48.47	48.05	47.41	46.95	46.60	46.32	46.10
5	47.18	37.12	33.20	31.09	29.75	28.83	28.16	27.65	27.24	26.92	26.42	26.06	25.78	25.57	25.39
6	35.51	27.00	23.70	21.92	20.80	20.03	19.46	19.03	18.69	18.41	17.99	17.68	17.45	17.27	17.12
7	29.25	21.69	18.77	17.20	16.21	15.52	15.02	14.63	14.33	14.08	13.71	13.43	13.23	13.06	12.93
8	25.41	18.49	15.83	14.39	13.48	12.86	12.40	12.05	11.77	11.54	11.19	10.94	10.75	10.60	10.48
9	22.86	16.39	13.90	12.56	11.71	11.13	10.70	10.37	10.11	9.89	9.57	9.33	9.15	9.01	8.90
10	21.04	14.91	12.55	11.28	10.48	9.93	9.52	9.20	8.96	8.75	8.45	8.22	8.05	7.91	7.80

(continued)

Table A.3 (continued)

	v_1														
v_2	1	2	3	4	5	6	7	8	9	10	12	14	16	18	20
11	19.69	13.81	11.56	10.35	9.58	9.05	8.66	8.35	8.12	7.92	7.63	7.41	7.24	7.11	7.01
12	18.64	12.97	10.80	9.63	8.89	8.38	8.00	7.71	7.48	7.29	7.00	6.79	6.63	6.51	6.40
13	17.82	12.31	10.21	9.07	8.35	7.86	7.49	7.21	6.98	6.80	6.52	6.31	6.16	6.03	5.93
14	17.14	11.78	9.73	8.62	7.92	7.44	7.08	6.80	6.58	6.40	6.13	5.93	5.78	5.66	5.56
15	16.59	11.34	9.34	8.25	7.57	7.09	6.74	6.47	6.26	6.08	5.81	5.62	5.46	5.35	5.25
16	16.12	10.97	9.01	7.94	7.27	6.80	6.46	6.19	5.98	5.81	5.55	5.35	5.20	5.09	4.99
17	15.72	10.66	8.73	7.68	7.02	6.56	6.22	5.96	5.75	5.58	5.32	5.13	4.99	4.87	4.78
18	15.38	10.39	8.49	7.46	6.81	6.35	6.02	5.76	5.56	5.39	5.13	4.94	4.80	4.68	4.59
19	15.08	10.16	8.28	7.27	6.62	6.18	5.85	5.59	5.39	5.22	4.97	4.78	4.64	4.52	4.43
20	14.82	9.95	8.10	7.10	6.46	6.02	5.69	5.44	5.24	5.08	4.82	4.64	4.49	4.38	4.29
21	14.59	9.77	7.94	6.95	6.32	5.88	5.56	5.31	5.11	4.95	4.70	4.51	4.37	4.26	4.17
22	14.38	9.61	7.80	6.81	6.19	5.76	5.44	5.19	4.99	4.83	4.58	4.40	4.26	4.15	4.06
23	14.20	9.47	7.67	6.70	6.08	5.65	5.33	5.09	4.89	4.73	4.48	4.30	4.16	4.05	3.96
24	14.03	9.34	7.55	6.59	5.98	5.55	5.23	4.99	4.80	4.64	4.39	4.21	4.07	3.96	3.87
25	13.88	9.22	7.45	6.49	5.89	5.46	5.15	4.91	4.71	4.56	4.31	4.13	3.99	3.88	3.79
26	13.74	9.12	7.36	6.41	5.80	5.38	5.07	4.83	4.64	4.48	4.24	4.06	3.92	3.81	3.72
27	13.61	9.02	7.27	6.33	5.73	5.31	5.00	4.76	4.57	4.41	4.17	3.99	3.86	3.75	3.66
28	13.50	8.93	7.19	6.25	5.66	5.24	4.93	4.69	4.50	4.35	4.11	3.93	3.80	3.69	3.60
29	13.39	8.85	7.12	6.19	5.59	5.18	4.87	4.64	4.45	4.29	4.05	3.88	3.74	3.63	3.54

(continued)

Table A.3 (continued)

	v_1														
v_2	1	2	3	4	5	6	7	8	9	10	12	14	16	18	20
30	13.29	8.77	7.05	6.12	5.53	5.12	4.82	4.58	4.39	4.24	4.00	3.82	3.69	3.58	3.49
35	12.90	8.47	6.79	5.88	5.30	4.89	4.59	4.36	4.18	4.03	3.79	3.62	3.48	3.38	3.29
40	12.61	8.25	6.59	5.70	5.13	4.73	4.44	4.21	4.02	3.87	3.64	3.47	3.34	3.23	3.14
50	12.22	7.96	6.34	5.46	4.90	4.51	4.22	4.00	3.82	3.67	3.44	3.27	3.41	3.04	2.95
60	11.97	7.77	6.17	5.31	4.76	4.37	4.09	3.86	3.69	3.54	3.32	3.15	3.02	2.91	2.83
70	11.80	7.64	6.06	5.20	4.66	4.28	3.99	3.77	3.60	3.45	3.23	3.06	2.93	2.83	2.74
80	11.67	7.54	5.97	5.12	4.58	4.20	3.92	3.70	3.53	3.39	3.16	3.00	2.87	2.76	2.68
90	11.57	7.47	5.91	5.06	4.53	4.15	3.87	3.65	3.48	3.34	3.11	2.95	2.82	2.71	2.63
100	11.50	7.41	5.86	5.02	4.48	4.11	3.83	3.61	3.44	3.30	3.07	2.91	2.78	2.68	2.59
120	11.38	7.32	5.78	4.95	4.42	4.04	3.77	3.55	3.38	3.24	3.02	2.85	2.72	2.62	2.53
150	11.27	7.24	5.71	4.88	4.35	3.98	3.71	3.49	3.32	3.18	2.96	2.80	2.67	2.56	2.48
200	11.15	7.15	5.63	4.81	4.29	3.92	3.65	3.43	3.26	3.12	2.90	2.74	2.61	2.51	2.42
250	11.09	7.10	5.59	4.77	4.25	3.88	3.61	3.40	3.23	3.09	2.87	2.71	2.58	2.48	2.39
300	11.04	7.07	5.56	4.75	4.22	3.86	3.59	3.38	3.21	3.07	2.85	2.69	2.56	2.46	2.37
400	10.99	7.03	5.53	4.71	4.19	3.83	3.56	3.35	3.18	3.04	2.82	2.66	2.53	2.43	2.34
500	10.96	7.00	5.51	4.69	4.18	3.81	3.54	3.33	3.16	3.02	2.81	2.64	2.52	2.41	2.33
600	10.94	6.99	5.49	4.68	4.16	3.80	3.53	3.32	3.15	3.01	2.80	2.63	2.51	2.40	2.32
750	10.91	6.97	5.48	4.67	4.15	3.79	3.52	3.31	3.14	3.00	2.78	2.62	2.49	2.39	2.31
1000	10.89	6.96	5.46	4.65	4.14	3.78	3.51	3.30	3.13	2.99	2.77	2.61	2.48	2.38	2.30

Table A.4 Chi-square distribution table

d.f.	0.995	0.99	0.975	0.95	0.9	0.1	0.05	0.025	0.01
1	0.00	0.00	0.00	0.00	0.02	2.71	3.84	5.02	6.63
2	0.01	0.02	0.05	0.10	0.21	4.61	5.99	7.38	9.21
3	0.07	0.11	0.22	0.35	0.58	6.25	7.81	9.35	11.34
4	0.21	0.30	0.48	0.71	1.06	7.78	9.49	11.14	13.28
5	0.41	0.55	0.83	1.15	1.61	9.24	11.07	12.83	15.09
6	0.68	0.87	1.24	1.64	2.20	10.64	12.59	14.45	16.81
7	0.99	1.24	1.69	2.17	2.83	12.02	14.07	16.01	18.48
8	1.34	1.65	2.18	2.73	3.49	13.36	15.51	17.53	20.09
9	1.73	2.09	2.70	3.33	4.17	14.68	16.92	19.02	21.67
10	2.16	2.56	3.25	3.94	4.87	15.99	18.31	20.48	23.21
11	2.60	3.05	3.82	4.57	5.58	17.28	19.68	21.92	24.72
12	3.07	3.57	4.40	5.23	6.30	18.55	21.03	23.34	26.22
13	3.57	4.11	5.01	5.89	7.04	19.81	22.36	24.74	27.69
14	4.07	4.66	5.63	6.57	7.79	21.06	23.68	26.12	29.14
15	4.60	5.23	6.26	7.26	8.55	22.31	25.00	27.49	30.58
16	5.14	5.81	6.91	7.96	9.31	23.54	26.30	28.85	32.00
17	5.70	6.41	7.56	8.67	10.09	24.77	27.59	30.19	33.41
18	6.26	7.01	8.23	9.39	10.86	25.99	28.87	31.53	34.81
19	6.84	7.63	8.91	10.12	11.65	27.20	30.14	32.85	36.19
20	7.43	8.26	9.59	10.85	12.44	28.41	31.41	34.17	37.57
22	8.64	9.54	10.98	12.34	14.04	30.81	33.92	36.78	40.29
24	9.89	10.86	12.40	13.85	15.66	33.20	36.42	39.36	42.98
26	11.16	12.20	13.84	15.38	17.29	35.56	38.89	41.92	45.64
28	12.46	13.56	15.31	16.93	18.94	37.92	41.34	44.46	48.28
30	13.79	14.95	16.79	18.49	20.60	40.26	43.77	46.98	50.89
32	15.13	16.36	18.29	20.07	22.27	42.58	46.19	49.48	53.49
34	16.50	17.79	19.81	21.66	23.95	44.90	48.60	51.97	56.06
38	19.29	20.69	22.88	24.88	27.34	49.51	53.38	56.90	61.16
42	22.14	23.65	26.00	28.14	30.77	54.09	58.12	61.78	66.21
46	25.04	26.66	29.16	31.44	34.22	58.64	62.83	66.62	71.20
50	27.99	29.71	32.36	34.76	37.69	63.17	67.50	71.42	76.15
55	31.73	33.57	36.40	38.96	42.06	68.80	73.31	77.38	82.29
60	35.53	37.48	40.48	43.19	46.46	74.40	79.08	83.30	88.38
65	39.38	41.44	44.60	47.45	50.88	79.97	84.82	89.18	94.42
70	43.28	45.44	48.76	51.74	55.33	85.53	90.53	95.02	100.43
75	47.21	49.48	52.94	56.05	59.79	91.06	96.22	100.84	106.39

(continued)

Table A.4 (continued)

d.f.	0.995	0.99	0.975	0.95	0.9	0.1	0.05	0.025	0.01
80	51.17	53.54	57.15	60.39	64.28	96.58	101.88	106.63	112.33
85	55.17	57.63	61.39	64.75	68.78	102.08	107.52	112.39	118.24
90	59.20	61.75	65.65	69.13	73.29	107.57	113.15	118.14	124.12
95	63.25	65.90	69.92	73.52	77.82	113.04	118.75	123.86	129.97
100	67.33	70.06	74.22	77.93	82.36	118.50	124.34	129.56	135.81

Printed in the United States
by Baker & Taylor Publisher Services